'The study is informed by Brereton's vast expertise and continuing interest in environmental ethics and the public sphere. Consequently, the questions about audience-hood and environmentalism that he tackles are not limited to merely how media audiences consume and understand environmental messages, but rather how those audiences might actually be motivated to take social action. This makes his examination of the role of media in the formation of environmental citizenship, understood as civic participation in an age of ecological crisis, a crucial resource for those of us who are committed to confronting the challenge of climate change.'

Patrick D. Murphy, Associate Dean for Research and Graduate Studies, Klein College of Media and Communication, Temple University, USA

Environmental Literacy and New Digital Audiences

Environmental literacy and education is not simply a top-down process of disseminating correct attitudes, values and beliefs. Rather, it is one that incorporates and facilitates a dialogue with audiences of different persuasions and at all levels of engagement, to help highlight and co-produce consensual solutions to the major eco-challenges of our time.

Exploring the growing power and influence of media formats and outlets like YouTube and gaming, alongside fictional and documentary film, this book considers new modes of environmental literacy to ascertain the effectiveness of digital and filmic stimuli on an audience's perception of environmental issues, and its specific impact on environmental action. Drawing on extensive research across a broad range of media formats, Brereton establishes how environmental narratives and meanings are created and being received by contemporary audiences.

This book will be of great interest to students and scholars of environmental communication and media, eco-criticism and environmental humanities more broadly.

Pat Brereton is a Professor at Dublin City University, Ireland, and has taught and published extensively around environmental communications and all forms of new media.

Routledge Studies in Environmental Communication and Media

Environmental Communication and Critical Coastal Policy
Communities, Culture and Nature
Kerrie Foxwell-Norton

Climate Change and Post-Political Communication
Media, Emotion and Environmental Advocacy
Philip Hammond

The Discourses of Environmental Collapse
Imagining the end
Edited by Alison E. Vogelaar, Brack W. Hale and Alexandra Peat

Environmental Management of the Media
Policy, Industry, Practice
Pietari Kääpä

Participatory Networks and the Environment
The BGreen Project in the US and Bangladesh
Fadia Hasan

Participatory Media in Environmental Communication
Engaging Communities in the Periphery
Usha Harris

Journalism, Politics, and the Dakota Access Pipeline
Standing Rock and the Framing of Injustice
Ellen Moore

Environmental Literacy and New Digital Audiences
Pat Brereton

For more information about this series, please visit: www.routledge.com/Routledge-Studies-in-Environmental-Communication-and-Media/book-series/RSECM

Environmental Literacy and New Digital Audiences

Pat Brereton

Routledge
Taylor & Francis Group
LONDON AND NEW YORK

First published 2019
by Routledge
2 Park Square, Milton Park, Abingdon, Oxon OX14 4RN

and by Routledge
52 Vanderbilt Avenue, New York, NY 10017

First issued in paperback 2020

Routledge is an imprint of the Taylor & Francis Group, an informa business

British Library Cataloguing-in-Publication Data
A catalogue record for this book is available from the British Library

Library of Congress Cataloging-in-Publication Data
A catalog record has been requested for this book

ISBN 13: 978-0-367-66388-9 (pbk)
ISBN 13: 978-1-138-04925-3 (hbk)

Typeset in Times New Roman
by Wearset Ltd, Boldon, Tyne and Wear

To my youngest son Conor, who first told me about environmental YouTube videos, together with Rita and Robert who kept me grounded and gave me lots of ideas and criticism. Most sincerely to my wife Angela, for her constant support and consideration. Finally, to our dear colleague Louise McDermott who was the soul of Dublin City University (DCU) and left this life far too soon.

Contents

Acknowledgements

This book received financial support from the Faculty of Humanities and Social Sciences Book Publication Scheme at Dublin City University. Permission has been sought and gratefully received for re-framing an earlier version of the chapter on food documentaries, which appears in the *French Journal for Media Research* (2018).

I would like to thank the Faculty of Humanities of Social Sciences and the Executive Dean Prof. John Doyle for supporting my Sabbatical to write this book, having served as the Head of School of Communications for a number of years. Several staff and PhD students, especially from my own school, have been extremely helpful towards developing this ever-expanding area of research.

For my research trips abroad, I must thank among others, Brigitte Le Juez (Paris), Tommy Simpson (Berlin), Paul Smith and Chris Lawlor (short tour of a number of American Universities), Alison Anderson (Plymouth UK), Cristina Nistor and Elena Abrudan (Erasmus + trip to Cluj in Romania), Michael Roth who invited me onto a COST network focused on landscape and alternative energy, together with numerous trips to the UK, Spain, France, Scotland, Germany, Austria, the Czech Republic and other regions across the world. All of these journeys have helped to widen my understanding of environmentalism and new modes of literacy that have been invaluable for my research.

I would also like to acknowledge the Green Foundation, which facilitated some fascinating environmental meetings, *An Taisce* who trained me as a Climate Change Ambassador, The Royal Irish Academy where I have been co-opted onto their Environmental committee, together with ECREA and IAMCR academic Communications research networks and especially Anders Hansen and Pieter Maeseele for progressing the field of environmental communications internationally through these academic networks. At the Free University, I met Kirsten Jorgensen who give me a good understanding of German environmentalism, together with a committed green activist Katie Griggs who is an inspiration. In Paris, among other scholars I met Tommaso Venturini who talked of the power of media, while Leda Mansour was a breath of fresh air with her outsider's insights on France and the world.

Over the year of writing this book, I also had the great pleasure of contacting or meeting a number of exceptional academics from a wide range of institutions, who have been very inspiring. These include:

Ken Boyle (DIT Ireland), David Taffler (Muhlenberg University USA), David Wheatly (Cambridge University), Jeff Kelly Lowenstein (Grand Valley State University USA), John Barry (Queens University), John Parham (University of Worchester), Joni Adamson (ASU America), Lisa Garford (Newcastle UK), Patrick Murphy (Temple University Philadelphia), Peter Gross (Tennessee America), Pietari Kääpä (Warwick UK), Scott Slovic (Idaho America), Sean Phelan (Massey University Australia), Hunter Vaughan (Oakland USA), Alexa Weik von Mossner (Klagenfurt Austria) and many others besides. To my long-time proof reader, indexer and best friend, Michael Doorley.

1 Environmental literacy

An introduction

Overview

With a growing backlash towards climate change adaptation, greater emphasis has to be placed on promoting new modes of education and environmental literacy across all aspects of media and communications studies. At the outset, this study tends to endorse eco-literary scholar Greg Garrard's (2012) assertion that the definition of what counts as pollution, much less any major environmental issue, up to and including climate change, remains a cultural as well as a political and scientific question that depends as much on shifting values and priorities, as on actual emissions of toxic substances.

Environmental literacy and education is certainly not simply a top down process of disseminating correct attitudes, values and beliefs, but incorporates and facilitates a dialogue with audiences of different persuasions and at all levels of knowledge and engagement, to help highlight and at best co-produce consensual solutions to the major challenges of our time. Furthermore, it is affirmed that environmental education is not simply about 'saving the whale' or indeed saving local habitat's but is equally about the development of an appreciation of the wonders and beauty of the world and provoking a sense of *wanting* to save it. Consequently, as suggested in *Environmental Ethics and Film* (Brereton 2016), the development of a deeper form of ecological media thinking, while recognising all aspects of environmental justice, alongside promoting increased levels of critical literacy, is essential in creating a robust, sustainable environmental citizenship and embedding a comprehensive strategy for activating effective solutions into the future.

This study will go beyond an examination of fictional films to also include documentaries, as well as recognising the proliferation of YouTube, video games and other new media platforms for engagement with environmental issues. Discovering what might be considered as best practice with regards to such digital media outlets demands extensive engagement with the most popular and effective media artefacts. To begin this process of examining environmental communication opportunities afforded by the newer interactive media that are flourishing across a wide range of such social networks, it is necessary to tease out their psychological and ideological influences, alongside the unique affordances of various forms of media, and at the same time uncover potential tipping points for audience engagement and hopefully transformation. Later chapters will concentrate on YouTube as a new media platform, alongside a study of (serious) video games like *Pokémon Go*, which explicitly or implicitly address environmental issues, while helping to explore how they might assist in teaching and promoting critical environmental literacy (Rust *et al.* 2013). Such examples of transmedia and convergent formats in various ways speak directly to contemporary audiences and potentially invoke new modes of environmental literacy.

From media literacy to environmental literacy

Sonia Livingstone suggests that 'despite enthusiastic calls for new digital literacy programs' and the recent 'embedding of media literacy requirements within national and international regulation', there remains little agreement about media literacy or 'how to measure it, and therefore little evidence that efforts to improve it are effective' (Livingstone 2011). This assessment also rings true with regards to the development of environmental literacy.

If audio-visual media are to play a significant role in facilitating participation and the development of environmental education in the public sphere and towards promoting democratic values that cuts across inequalities at all levels, much needs to be teased out. In particular, far greater understanding of the power of mediated texts is needed, coupled with an exploration of how audiences consume such texts, together with an appreciation of such textual interaction with society, as they feed into a range of intersecting institutional tensions. Furthermore, policy around (environmental) literacy should be grounded in the experience of media use, learning, expression and civic participation among citizens, and at the same time should be developed and implemented through collaboration with academic stakeholders, schools, the media industry and civic society (see Livingstone 2011).

This form of productive and active literacy can also be extended and incorporated into a broad policy model to help develop new forms of environmental literacy and productive modes of eco-citizenship. Media literacy in general, and environmental literacy in particular, remains the most fruitful strategy to frame future media audience investigation and incorporate both an explanatory and a normative agenda, rather than provoking a divisive partisan and isolationist position, which unfortunately so much complex environmental debates tend to engender. Environmental scholars, I believe, should continue to go back to basics and ask well-established media questions, such as: what do citizens and consumers know about their changing media environment and, further, what should they know about the environment? And then, most critically, pedagogical analysis and curricular tools and protocols need to be explored to help tease out questions like: what does it matter if they don't have this knowledge and, in whose interest is it if they do?

Methodological strategies

Alongside well-established textual analysis protocols used to examine film from an environmental and educational perspective (see Brereton 2005; Rust *et al.* 2016), such strategies will be re-applied and extended towards exploring a wide range of online media. For instance, later chapters on streaming Netflix televisual series, YouTube artefacts and environmentally predisposed video games will serve to extend and even problematise conventional feature length film analysis strategies around environmentalism, together with addressing eco-documentaries and other new media formats. All of these cross-platform and converging media outlets are being extensively colonised and re-appropriated by growing new-generational audiences in particular. Reception and psychological studies of audience behaviour and trigger points that address environmental issues will be explored in Chapter 2. For instance, highly tuned 'uses and gratification' and other research tools for audience engagement have been successfully used across so many contemporary studies in striving to address the inter-connection between mediated texts and audiences and these can be re-applied to explore new forms of online media, while capturing their potential for developing effective modes of environmental literacy.

Furthermore, the study will tentatively suggest how audiences might tease out meanings and interpretations, alongside exploring how new forms of literacy might evolve. Yet, as evident for instance through large-scale Eurobarometer surveys over the years, it remains difficult to gauge the effectiveness of filmic, much less other media stimuli on audience's perceptions of environmental issues (see Nisbet *et al.* 2018). While it remains difficult to tease out such concerns, at the same time there is a danger of harking back to early effects debates (see Payne Fund research in the USA from the 1920s onwards) and its reductive assumptions around hypnotic media and passive audiences. The literature and methodological tools have thankfully become much more sophisticated of late, taking on broad notions of active audiences, alongside empirically applied psychological methodologies of investigation around human behaviour, as outlined in Chapter 2.

In essence, this study will strive to achieve its limited goals[1] by:

1 *Criticising* and analysing new digital online artefacts, together with a corpus of popular films, documentaries and televisual series that can ostensibly provoke a range of environmental issues and debates.
2 *Applying* the literature used in traditional audience and reception studies to tease out what might be considered the most effective tipping points and especially highlighting particular narrative stimuli in developing effective environmental media results.
3 *Addressing* the textual analysis protocols and potential reception of environmentally focused productions, delivered through online and so-called legacy media.
4 *Identifying* specific content and production possibilities that serve to enhance audience responses and most especially promote environmental learning protocols.

Environmental literacy as a new mode of learning

The concept of environmental literacy first appeared in a 1969 article by Charles Roth, in response to the then frequent American media references to environmental illiterates who were polluting the environment. Little attention was paid to the essay until a year later, when President Richard Nixon used it in relation to the passage of the first National Environmental Education Act.

Environmental literacy as a goal of general education is constituted by a number of precepts which include:

- all sustainable human activities are dependent upon a clean, healthy and productive environment
- it is the environment that provides the materials and energy to meet our basic needs and desires
- the nature of particular environments sets parameters for many human activities and establishes risks for those activities
- all human activities have consequences for the environment both positive and negative
- Environmental literacy is essentially the capacity to perceive and interpret the relative health of environmental systems and take appropriate action to maintain, restore, or improve the health of those systems.

(Roth 1968: 10)

Formal environmental education began to emerge as a distinct field in the mid-1960s. It has its roots in a variety of related fields, including conservation education, nature education, resource-use education, outdoor education, geographic education and of course science education. It draws its strength from an examination of several macro issues including:

- the interrelationships between natural and social systems
- the unity of mankind with nature
- technology and the making of choices
- developmental learning throughout the human life cycle.

(Roth 1992: 17)

Although the pedagogical and communications relationship between cognitive components, affective components and overall behaviour is complex and not necessarily linear, researchers have ostensibly shown that increasing an individual's environmental knowledge through study results in more positive attitudes towards the environment (Bradley *et al.* 1999) and in turn can help to create more responsible environmental behaviour. While a number of caveats must be added to such a broad ranging assertion, most especially the pervasive need for large scale systematic change, nonetheless this still remains the general goal of many environmental programmes. Knowledge of course remains a critical component of environmental literacy. But as we discover this on its own is not a sufficient precursor for responsible environmental behaviour (see Nisbet *et al.* 2018).

In particular, the engagement and valuing of personal agency has become essential as a prerequisite, while the environmental problem grows exponentially more acute and by all accounts remains beyond the scope of individual actions. As will be constantly affirmed, heroic narrative engagement across environmental concerns is needed to serve as a constant reminder of such potentiality. Nonetheless, apathy and various forms of fatalism often seep in very quickly when addressing the broad sweep of environmental issues. As conceptualised by social scientists, one of the tools developed to measure environmental attitudes is the New Environmental Paradigm (NEP) scale, developed by Dunlap and Van Liere and later revised as the New Ecological Paradigm Scale (see Dunlap 2008) to address contemporary issues in the environmental debate.

The NEP represents an ecocentric (pro-environmental) worldview that among other aspects, recognises the intrinsic value of nature (i.e. anti-anthropocentrism), the fragility of nature's balance, limits to economic growth, and most importantly the possibility of future ecological crisis. Such a procedure allows researchers to track whether values and attitudes towards nature are changing over time and to gauge the contribution of environmental education to changing environmental attitudes (Peer *et al.* 2007: 47). As this book highlights, long-term cross-national audience research is needed to constantly test and access the potency of mass mediated texts. While this is the most effective strategy and lens to use, this remains a broad-church, inclusive framework of environmentalism.

Media literacy has become one of the key qualifications for taking an active part in contemporary society. As media technology apparently becomes more intuitive and media and other social practices intertwine more and more, this book uses numerous aspects of literacy to better understand what makes someone media literate, as well as how this influences social development. Media literacy has been effectively defined as the ability 'to access, analyse, evaluate and communicate messages in a variety of forms' (Aufderheide and Firestone 1993: xx). Other scholars highlight the concept of information literacy (Koltay 2011), digital literacy (Hobbs 2011), ICT literacy (Friemel and Signer 2010), new media literacies (Jenkins 2006) and most recently, social media literacy (Livingstone 2016). Today communication scholars stress that both the media's offerings, coupled with social interactions play a crucial role in understanding an ever-evolving mode of (digital) media literacy (Livingstone 2016). Following theoretical considerations around developing robust models of democracy, media literacy has a primary emphasis placed on 'knowledge' (Potter 2010: 680), 'information' (Buckingham 2007: 45), and 'analytical competencies' (Koltay 2011: 217), all the while great emphasis is generally placed on the development of critical thinking.

Sonia Livingstone (2016) further suggests that the problem of all forms of illiteracy rests less on awareness and more with acting upon that knowledge. This conundrum certainly also applies to environmental education and communication.

Media knowledge per se does not necessarily lead directly to a critical evaluation of media content, much less to a potential change in behaviour (Martens 2010: 13). To help resolve this apparent disconnect, through future research and therefore to answer the quintessential 'so what' question, scholars need to proceed several steps further. Most particularly, in asking what are the 'steps you can take to act on your knowledge' (Silverblatt *et al.* 2009: 106). Essentially, research needs to look at the social process involved in normalising and promoting media literacy and by extension, for the purposes of this study, promoting new forms of environmental literacy.

To help integrate such educational and social norms successfully, individuals have to 1) perceive others' beliefs about a given behaviour and 2) be motivated to comply with those beliefs (Lapinski and Rimal 2005). Deci and Ryan (2000) offer a growth perspective on development and education that allows us to analyse how media literacy can 'promote personal discovery and growth' (Silverblatt *et al.* 2009: 77). Meanwhile, self-determination theory empirically identifies three basic needs which are necessary for such growth and well-being: autonomy, relatedness, and competence. Addressing all of these needs and approaches to media knowledge provide a benchmark and precondition for being able to benefit from the media, while avoiding some inherent risks (Hobbs 2011: 421).

In summary, one would have to conclude that the most important literary skill for audiences to acquire is the ability to evaluate media content and to reflect upon the 'impact of media messages and technology upon our thinking and actions in daily life' (Hobbs 2011: 12). Rather than simply developing critical engagement, which is often read as the only clear benchmark of effective literacy, it is recognised that much of the time spent on internet communication in particular is simply 'used to maintain and deepen existing friendships, which eventually enhances the closeness of these friendships' (Peter and Valkenburg 2013: 687).[2] While not in any way demeaning towards such everyday social usage, this does not appear to support more active and reflexive forms of literacy, including for that matter environmental engagement. Yet much of the scholarly literature, as evident across the subsequent chapters seems to unconditionally infer that new media affordances de facto promote increased transmedia and active audience engagement.

Active audiences, new media and multimodality

While highlighting the production of media texts – as well as producing bottom-up textual analysis studies – it is helpful to apply a diagnostic critique to help reveal Hollywood and other new media's implicit messages concerning a number of ecological issues. Kellner (1995) most notably suggests that this method's purpose is designed to foster critical evaluation of key ideological messages emanating from mainstream media culture. In terms of textual analysis, Kellner advises that one must carefully evaluate the production of media culture to assess latent meaning through recognition of various meta-narratives, myths and symbols in order to discern emerging patterns overall. For Kellner (1995: 98) 'a multiperspectival cultural studies draws on a wide range of textual and critical strategies to interpret, criticize and deconstruct the artefact under scrutiny.' This is especially true for environmental work, which has long used a diversity of critical models 'as a quest for adequate models of inquiry from the plethora of possible alternatives that offer themselves from whatever disciplinary quarter' (Buell 2005: 10).

One of course cannot understand commercial popular culture in any meaningful way, without as Hall (1981: 231) and many other scholars advocate, 'taking into account the monopolization of the cultural industries'. As a constant flow of studies testifies, when it comes to consideration of political economy, it is difficult to overstate just how concentrated the media landscape has become. As an environmental media scholar, who remains preoccupied with recuperative textual analysis, one cannot stress this tension enough, especially as a counterweight to ongoing attempts to privilege a cultural studies and audience reception approach to popular media. One further wonders, however, if such concerns are even more fraught and complicated by the expediential growth of new digital media formats like streaming TV, YouTube and video games, etc.

This study will nonetheless concentrate on mainstream and so-called legacy media, together with the proliferation of online web material, while ostensibly attempting to raise and present narrative solutions, often 'against the grain' (or through oppositional readings, as discussed in subsequent chapters) to environmental problems. All the while, there remains a pervasive danger of simply overstating the utopian potential of new digital media in particular of provoking much needed change and transformation. See for instance Timothy Clark's (2015) *Ecocriticism on the Edge*, which identifies some of the dangers of being too 'positive' with regards to eco-cinema analysis and warns scholars against being 'closet apologists for global capitalism' (11). This cautioning approach, serves as a necessary counterweight to the apparent 'blind optimism' around the potential positive power of media – especially new digital media – in helping transform environmental citizenship, towards actively addressing the global environmental crisis on our planet, while at the same time embracing all forms of consumption as empowering! This latent paradox, embedded within the commercial media industries *modus operandi*, cannot be ignored in the face of a global ecological crisis.

Lisa Garforth, in her latest study, *Green Utopias* (2018), provides a very balanced evaluation of the utopian potential of literature and by extension media by concluding:

> We are not greeting the Anthropocene with only hope, only fear, only apocalypse, only adaptation. If we have learned nothing else about our environmental predicament since the limit to growth, it is that it is huge in physical and conceptual scale, it is diverse in content, and it is thoroughly (as Hulme puts it) wicked: multi-dimensional and essentially unresolvable. There is no single or elegant solution. There is no one way of imagining a better post-carbon future. [But] [t]here is always hope.
>
> (162)

So the power of media has to be appropriated in whatever way can be found, all the while recognising the risks involved.

Furthermore, it is important to note when conducting textual analysis which is not explicitly paired with direct (empirical) studies of audience reception, that such analysis can easily run the risk of veering too closely to what Kellner refers to as 'manipulation theory' (cited in Moore 2017: 18). Most specifically, one wants to avoid conceptualising the (imagined) audience, like Stuart Hall's 'blank screens', who are in turn easily dominated by mass culture and by the multinational media incorporations that have a guiding hand in creating such culture.

Alternatively, this study supports the stance that the relationship embedded within the media landscape could be considered as representing a complex interconnection of society and media. Some describes this as one that avoids conceptualising influence in only one direction. To further avoid these polarising extremes, Kellner refers to the need for a 'dual optic', which identifies both 'utopian' and more pessimistic 'ideological' perspectives across media culture (in 1995: 93). In so doing, one could potentially re-construct environmental literacy as a project which attempts to incorporate the notion of popular culture as a site of transformation and at the same time recognise commercial popular culture as an 'arena of both consent and resistance, a contested zone of meaning', as Litzinger (2001: 254) insightfully identifies.

Essentially, as eco-film scholars Rust, Monani and Cubitt (2013) note, cinema and media in general 'is a form of negotiation, a mediation that is ecologically placed, as it consumes the tangled world around it' (1). At the same time, one should further caution that by any scientific measure there is not enough explicitly environmentally influenced mainstream (Hollywood) narratives for audiences to consume, much less new media artefacts to draw from, in the struggle towards developing a robust corpus of environmentally media texts. In itself this dearth of material might infer an ideological stifling of such globally important issues but, as examined across the next few chapters, there are also many psychological, as well as ideological barriers for such a vacuum being created and maintained, in spite of an expediential growth of media output. By any measure there remains a lack of environmentalism being registered or even tangentially evident across the global communications political sphere. Certainly, this is a major shortcoming[3] and signals the abiding need for creatives and environmental activists to proactively assist in developing more explicit and effective audio-visual media projects, which cogently speak to the apocalyptic climate change situation.

In spite of all these caveats however, some more hopeful signs can be recognised within the growth of new online media, affording a corpus of provocative storylines that speak most directly to new generational tastes and interests. This volume will flesh out some of these, including the rapper Prince Ea with his powerfully evocative climate change music video *Sorry* discussed in Chapter 10. But one would have to agree that much more diverse output is needed by the media industries to create original and effective narratives that broaden and help to deepen the audio-visual landscape. This is needed to support the growth of environmental literacy in the future and to help keep climate change on the agenda, which remains the single most important challenge of our age.

All forms of new screen media have in many ways become multimodal, which can appeal to environmental concerns, yet the visual still remains prioritised. In a growing music-saturated contemporary world, such sensory dominance, with its fast-paced modes of interactivity, is challenged. Neil Postman's worries over losing a 'literate culture' in *Amusing Ourselves to Death* (1985) have certainly reached a new level of anxiety, with the mushrooming of videogaming and other forms of new media consumption. But such multi-layered affordances, coupled with a legacy that media provide increased opportunities for developing powerful immersive narratives can alternatively speak to the growing complexity of environmental issues. Certainly, digital games have sought to combine technological newness of form with a strong link to the familiar, while at the same time privileging popular cultural narratives and immersive meanings (Brereton 2012: 209).

The following summarises some of the key transformations facilitated by new media innovations, as suggested by a broad range of scholarly research (see Metzger in Nabi and Oliver 2009). These include:

1. Interactivity – Web 2.0 including blogs and YouTube illustrates the power of interactivity and opens up possibilities for audiences to affect media content.
2. The range of content has increased, becoming multimodal, consequently media effects and active textual engagement with a less consistent set of messages has radically changed audience engagement.
3. Audiences can be more selective and can easily time-shift their consumption of media in various ways in the move towards narrowcasting.
4. Audiences can constantly tailor media to suit their personal needs and wishes.
5. Increased convergence and the blurred boundaries and multiple modes evident across so much media output is becoming normative and mainstreamed.
6. Hypertextuality and the global reach of new media is ever-increasing.

All of these attributes and affordances will be explored in varying ways across subsequent chapters. New media most certainly improves the ability to extend social connectivity across their audiences, which is important because the social interaction surrounding mass media exposure can influence viewers interpretation of media content and thus influence how that content may affect them (Metzger in Nabi and Oliver 2009: 566). This form of multimodality can probably influence the growth of critical forms of environmental literacy.

Transforming audiences across (new) media and new forms of environmental literacy

Jenkins (2006) locates the specificity of present-day media cultures in the combination of top-down business with bottom-up consumption and production practices. Bottom up audience analysis has become the primary avenue for research of late. For much of its history, the study of media audiences has focussed primarily on what people *do* with the media, how they read, interpret and respond to media texts, while focusing on the impact or effects that media may have on audiences as a result of such interactions. A secondary interest has been focused on the actual social relationships, which result from being part of an audience. Here the dynamics of audience agency is key and in accepting participants as social actors, audiences use 'media to suit themselves' (Webster 1998: 194).

For example, see Rosen's influential essay 'the people formally known as the audience', where he argues that (commercial) media systems have lost (ideological) control over its audiences, who have been (re)transformed into 'the public made realer, less fictional, more able, less predictable' (2008: 165). Among other reasons, this is why Denis McQuail talks of the hybridity of the audience and the rise of the 'user' in digital media studies. Lievrouw and Livingstone (2002) also effectively assert that 'there is an uncertainty over how to label people in terms of their relationship with new media and how in a number of important ways audiences have become simply characterised as "users"'. Ostensibly, it appears best to explain online audience activity as a phenomena where people are perceived to 'use' media technologies and content more actively.

By actively reading 'against the grain', this notion of the user can help further uncover more fruitful environmental lessons. Even conventionally-driven new media can suggest such a productive form of interactivity, as illustrated through many surprisingly fruitful eco-tales discussed in subsequent chapters. All the while, of course, it must be recognised that there continues to be a surprisingly poor diet of overt, much less covertly conceived environmental narratives. This famine belies the global importance and gravity of the problem. The 'advertising revenues that sites such as YouTube, Facebook and MySpace generate, are derived substantially from audience attention captures, with content produced by members of the user/audience community' (Napoli 2010: 512).

Furthermore, all social relationships are now mediated in some form or other, leading some researchers to ask new questions, such as 'what are people doing that is related to media' (Couldry 2012). We can look at the evolution of audience 'prosumption' or 'produsage' (Bruns 2008) and at audience 'activism' (de Jong *et al.* 2005), as illustrating new dynamics of social interaction and self-representation, but with the potential to contribute to larger, integrative social networks that transcend the existing boundaries of the traditional conception of the audience. All of these oscillating debates and definitional tensions can be usefully ascribed to environmental literature protocols, as explored across later chapters.

Nightingale (2004) usefully identifies a four-fold typology within which audience experience intersects with social structures, positing publics, markets, communities and fandoms, as constituting the principal axes around which social relations are formed. Within the established audience research tradition, fandom and fan studies most successfully provides a unique area of enduring interest in social relationships created through processes of media reception, beginning with what Jenkins termed 'textual poachers'. Jenkins has called attention to the many rich forms of creative audience engagement with media content, such as appropriating and refashioning cultural artefacts and resources. More recently the term Web 2.1 or the 'participatory web' encompasses internet websites that allow users to interact and collaborate, hence 'participate' and such research trajectories tend to follow utopian appraisals, including Rhengold's (1993) *The Virtual Community* and Castells (1998) *Information age* trilogy. The twenty-first century, by all accounts, has witnessed a boom in interest in and work around the notion of the prosumer and the productive process of presumption (Ritzer *et al.* 2012). Yet at the same time one must always remember the relatively low penetration rate of truly active social media usage, as explored in subsequent chapters. Furthermore, from an audience studies perspective, it is important to look beyond the rhetoric of the over-celebratory approaches towards new media, which this study remains constantly cognisant of, as it weaves across a broad range of media platforms and outlets.

Post-politics and new religious modes of environmental engagement

We have only just begun to understand what the dynamics of climate change is doing to the planet. 'Not just what it is doing to the ecological and physical conditions of our existence but, more importantly, to our political discourses, social relationships and imaginative worlds' (Hulme 2010: 273–274). Others contend nothing much has changed. John Urry (2011: 91) more hopefully maintains that politics is alive and well in the grassroots and on the internet. He dismisses the post-political critique, as nostalgia for a 'golden age' of democratic engagement that never existed. Meanwhile Jodi Dean (2011) is even more dismissive claiming that post-politics is 'childishly petulant'. Effectively, the Left is saying 'if the game isn't played on our terms, we aren't going to play at all. We aren't even going to recognise that a game is being played' (in Hammond 2018: 4).

Climate change has become a defining bell weather debate across this post-political discourse. Indeed, for Swyngedouw (2013: 3), it is the 'emblematic case' and '*cause celebre* of de-politicization'. Hammond affirmed that a concern with the mainstreaming and de-politicisation of climate change is the common starting point for such debate. This is not necessarily seen as a problem inherent within green politics as such, but as something that has happened to it, or around it, and which has compromised its former radicalism (Hammond 2018: 6). Alternatively, as Chris Methmann (2010: 345) insightfully argues, 'climate protection' has become an 'empty signifier', adopted as an 'important policy goal' not only by national governments and the United States, but also by global economic institutions such as the World Trade Organisation, IMF, World Bank and OECD etc. (Hammond 2018: 6). Why would it appear necessary to 'integrate climate protection into the global hegemonic order, without changing the basic social structures of the world economy' (Methmann 2010: 345).

Furthermore, according to Swyngedouw (2013: 228), environmentalism now functions as a 'new opium of the masses' playing the role of forestalling various conflicting and alternative trajectories, while reinforcing a liberal-capitalist order for which there seems to be 'no alternative'. In making this argument, Swyngedouw follows the lead of Alain Badiou, who pronounced in a 2007 interview that 'the rise of the rights of Nature' is a contemporary form of the opium of the people.

> It is an only slightly camouflaged religion: the millenarian terror, concern for everyone to save the properly political destiny of people, new instruments for the control of everyday life, the obsession with hygiene, the fear of death and of catastrophes.
>
> (Badiou in Feltham 2008: 139)

These fears are often played out through 'end of the world' environmental parables as explored across several chapters in this study.

Incidentally, Jean Baudrillard originally commented in this vein back in the 1970s, suggesting that environmentalism was 'a new opium of the people' (1974), while considering various forms of environmentalism as a 'witch hunt', attempting to unite antagonistic social classes in a 'new crusade' against a mystified threat by 'shouting apocalypse'. Hopefully, green politics and environmental communication have progressed some way from such polarising positions, but maybe not as much as one might hope, as again alluded to across various chapters.[4]

More conventionally, from a mainstream environmental communications perspective, Anabella Carvalho (2010: 172) remains more optimistic that the news media in particular can play an important role in 'processes of political (dis)engagement in relation to climate change'. She focuses on the way that media representations 'construct particular subject positions' for individuals and cultivate dispositions to action or inaction'. Carvalho goes on to suggest that the media generally present climate change as the concern of elite decision-makers, which pushes ordinary citizens back to the role of 'spectators or bystanders' (Carvalho 2010: 174–176). Beyond news and politics, I suggest fictional media formats can possibly contest and re-frame such ideological polarising positions and help audiences find a way through the morass of so much pontificating around various perspectives, while learning how to critically engage with such multi-faceted environmental debate. To help illustrate some of these polarising tensions, later chapters will address new innovative forms of media and provide case studies which serve to highlight a range of semiotic tools that can be used for critical analysis, which alternatively can assist in highlighting environmental literacy protocols and strategies.

Some areas for further environmental research

In general, this study posits a broad range approach of areas that need to be addressed in the on-going attempts to promote effective environmental literacy. These areas include:

- The need to strengthen contextual and process-related approaches to media and environmental literacy, without neglecting technological requirements for engagement and possible transformation of media systems.
- Reinforcing life skills around the definition of media and environmental literacy. Although Livingstone (2004: 12) states that 'the literacy associated with the new media, especially the internet, differs significantly from that of print and audio-visual media, we would like to propose defining skills primarily by their necessity for social development (e.g. critical thinking, social, or moral skills) and then adapting these skills to specific media and environmental literacy developments.' This remains the most appropriate pedagogical strategy to follow.
- Compare socialisation agents across young and older audiences. Research to date has not compared all known socialisation agents from a process-related perspective. For example, who mediates which skills? How does one's own motivation, media knowledge, and 'the quality of family relationships, as well as family composition' influence the mediation process (Vandoninck *et al.* 2013: 75).
- Reflecting on research norms around environmental literacy remains a preoccupation. Because 'all communication involves ethical and social values' (Hobbs 2011: 17), researchers should reflect on their own norms and aims and on how these values influence their own research. Moving beyond advocating for environmental literacy as a discrete, normative body of educational delivery, towards a more discursive, contingent and evolving area of investigation, demands clear articulation of future demands for effective environmental literacy. This should be mapped across the full range of demands, while especially taking into account the specific needs and demands of the First and Third World environments and all habitats in between.

There can finally be no simple or single definition of media (environmental) literacy, while the historical and cultural complexities of the phenomenon make its examination all the more urgent for a research and policy agenda. It is also crucial to recognise the limits of media and environmental literacy, especially given the growing technological complexity and deregulatory regime shaping digital media (Lunt and Livingstone 2012).

If media and environmental literacy is to play a significant role in facilitating civic participation, promoting democratic values and supporting the development of the world's post-carbon economic future, then greater institutional and holistic educational support is required. The following chapters will illustrate how audience psychology and various mediated trigger points can help to develop this critical educational agenda, as we work through case studies across a broad range of media formats.

Overview of chapters

Chapter 2, 'Understanding audience psychology and trigger points for promoting environmental literacy', focuses on a broad range of psychological and behavioural theory which has been used to explain the lack of take-up of environmental agendas and the need for increased models of effective environmental literacy. Exploring how the media and communications strategies generally can assist in this process of addressing environmental issues, culminating in dealing with the complexities of climate change, remains an ongoing concern. More broadly, environmental literacy and engagement has historically been framed within the flowering of first wave environmentalism that embraced a deep love of nature, coupled with a more recent second wave of critical engagement with nature and environmentalism, taking into account concerns around global justice and the growing urbanisation of contemporary society. A recent Netflix series *Ozark* is used to help illustrate some of these evolving debates.

Chapter 3, 'Promoting new media literacy: engagement with sustainable environmental literacy', outlines the growing literature around new media and its affordances, which are believed to be more active and engaging than earlier so-called passive media formats. Addressing the pervasive dangers of greenwashing and simply using environmental issues and nature generally to sell more products – including food and agriculture – are set out and illustrated through a case study of Irish farming. Nevertheless, the chapter counter-argues that such pervasive forms of greenwashing can be turned on its head, towards promoting productive forms of critical and active environmental literacy.

Chapter 4, 'Food documentaries and green anxieties: actively promoting environmental literacy', focuses on probably the most prescient environmental media topic, which speaks to large audiences across the world. Some claim such food documentaries have been the stimulus for many young people in particular moving towards vegetarianism and other environmentally sensitive forms of sustainable consumption. As the global agricultural industry is becoming more and more industrialised, which powerful documentaries like *Food Inc.* and *Cowspiracy* attest to, they also highlight the growing cruelty of factory farming, while more art-house documentaries like *Our Daily Bread* serve to create an alternative aesthetic, which in turn privileges a counter-discourse, promoting a slow-food agenda.

Chapter 5, 'Eco-documentaries: old problems, new aesthetic opportunities', teases out how new modes of environmental documentary serve to 'hit the sweet spot' between preaching as against wallowing in pleasurable sensation for mass audiences. While drawing on a number of well-established behavioural theories, including nudging and appealing to PR and most especially celebrity endorsement, this chapter will set up two contrasting readings of what might be broadly characterised as explicit environmental documentaries, encapsulated by *An Inconvenient Sequel*, which is contrasted with a more light-hearted, magazine style, celebrity-driven approach evident in *James May's Big Ideas*.

Chapter 6, 'Contemporary Hollywood blockbusters and environmental narratives', draws on a small selection of commercially successful contemporary storylines to continue to illustrate how such tales can be read – often against the grain – as promoting a broad range of environmental agendas and themes. This chapter will explore how at least the trace of a deep environmental agenda continues to be evidenced across science fictional narratives such as *Passengers*, *Blade Runner 2049*, alongside more conventional generic tales like *Sully* and *Deep Water Horizon*.

Chapter 7, 'An environmental analysis of post 9/11 American television series', attempts to co-opt big-budget television series into a productive environmental media discourse. This is long overdue, as such quality televisual series have evolved to become the most innovative form of media developed over the last few decades. Focusing on two classic series, *Sopranos* and *Homeland* – which ostensibly have little to do with environmental concerns per se – this chapter strives to illustrate how their use of immersive and long-play character-driven storylines can help towards addressing a range of explicit environmental issues, including waste disposal, global surveillance and various forms of environmental injustice.

Chapter 8, 'Netflix and emerging networks: new forms of immersive and addictive narratives and characterisation', is a companion piece to the previous chapter and concentrates on the immersive and even 'binge viewing' pleasures of the Netflix platform. Global online audiences are encouraged to engage with long play dramatic series like *Breaking Bad* and *Better Call Saul* for many hours of rewarding engagement. One could almost argue that such new immersive series have become the new-generational form of Shakespearian quality drama, which speaks across many layers of contemporary living, including an evolving environmental sensibility register. By closely focusing on place, identity and most especially speaking through three-dimensional characterisation, viewers can directly transport themselves into the mindset of these nefarious protagonists, who often unconsciously address very current and pervasive environmental issues. These include various forms of waste disposal and environmental ethics of all types, as they try to survive in the high stakes' world of drugs and criminality.

Chapter 9, 'Video games and environmental learning: new modes of audience engagement' draws heavily on a case study of *Pokémon Go*, provides a quick overview of serious games research and how this can help to seed and even promote new forms of active environmental literacy. With the gaming industry both online and offline growing rapidly, environmental scholars are tapping into ways to support and develop new forms of active environmental literacy. Most especially with the growth of vlogging and celebrity game-play, Twitch.TV for instance encourages new forms of online performativity, alongside Machinima which further engenders now modes of actively engaging with mediated environmental issues.

Chapter 10, 'Going viral: YouTube and new forms of environmental literacy', rounds off the volume with another major new media transformation. The Pandora's box of delights embedded within the YouTube platform provides so much educational material and potential, with the self-promoting of environmental issues popping up all the time. Because of space considerations, the chapter only calls attention to the proliferation of environmental music videos, while using a case study of an American rapper Prince Ea, who speaks directly to a broad range of environmental issues and has been extremely influential with a new generational audience.

Chapter 11 concludes the volume by drawing out some salient features from these broad ranging and inter-connecting chapters, while at the same time highlighting some important pedagogical strategies towards developing new forms of environmental literacy, which is essential if the ever-expanding media industry is to address and support an acute environmental agenda. Most certainly, if environmental communication scholars are to carry out their important role in promoting new pedagogical forms of engagement and support the development of new 'creative imaginaries' that speak especially to audiences across the world, much more active engagement with all forms of media and extending content and textual analysis exemplars are badly needed. Such open-ended scholarly activity helps to address the multi-layered problems for the environment while honing down on the wicked difficulty of dealing with climate change. As illustrated throughout this volume the careful use of narrative (including factual) media can serve as a focus for disseminating environmental concerns and especially suggesting a road map for the transition to a low carbon future.

Notes

1 Such grounded research case studies that need to be carried out will further help to explore and tease out the full gamut of human psychology embedded within the textual narrative power of environmental audio-visual stimuli.

2 There are four types of extrinsic motivation:

- external regulation – fear of punishment or hope for rewards motivates actions
- introjection – fear, shame and acceptance motivate action
- identification – value of a behaviour or a norm, even if they do not like the behaviour
- integration – children can integrate the required norm because it fits with their own values and beliefs.

(Vansteenkiste and Ryan 2013)

3 Even in print and broadcasting news, less than 1% of the total output constitutes direct environmental stories, according to several studies. On the other hand, fictional audio-visual media continues to have an uphill battle in gaining traction, especially within the broader environmental science communications fraternity.

4 Hammond concludes his very barbed commentary, that there is 'nothing better than a touch of ecology and catastrophe to unite the social classes' (ibid.: 7). According to others, it is climate change itself that presents an existential challenge to the capitalist order. Methmann (2010: 369) for instance argues that 'climate mainstreaming' serves the goal of 'sustaining capitalism': environmentalism has adopted a way of 'remedying the dislocatory effects of climate change for hegemonic structures without changing

them' (Hammond 2018: 8). Regarding climate change, most critics have understood the problem in terms of a lack of consensus, with the media giving too much space to a diversity of conflicting views about the causes of and responses to global warming (for example, Baykoff and Baykoff 2007). But what if the problem was the opposite? What if the promotion of consensus and narrowing of debate was encouraging inaction and disengagement rather than promoting change? (Hammond 2018).

References

Aufderheide, P. and Firestone, C. *Media Literacy, A Report of the National Leadership Conference on Media Literacy*. Washington DC: The Aspen Institute. 1993.

Baudrillard, J. 'The Environmental Witch-Hunt. Statement by the French Group 1970', in Reyner Banham (ed.) *The Aspen Papers: Twenty Years of Design Theory from the International Design Conference in Aspen*. New York: Praeger. Available online at: www.metamute.org/editorial/articles/environmental-witch-hunt.

Baykoff, M. and Baykoff J. 'Climate Change and Journalistic Norms: A Case-study of US Mass-media Coverage'. *Geoform*. 2007. Available online at www.eci.ox.ac.uk/publications/downloads/boykoff07-geoforum.pdf.

Bradley, J. C., Waliczek, T. M. and Zajicek, J. M. 'Relationship between Environmental Knowledge and Environmental Attitudes of High School Students', *The Journal of Environmental Education*, 30(3): 17–21. 1999.

Brereton, P. *Environmental Ethics and Film*. London: Routledge 2016.

Brereton, P. *Hollywood Utopia: Ecology in Contemporary American Cinema*. Bristol UK: Intellect Press 2005.

Brereton, P. *Smart Cinema: DVD Add-ons and New Digital Audiences*. Basingstoke UK: Palgrave 2012.

Bruns, A. *Blogs, Wikipedia, Second Life and Beyond: From Production to Produsage*. New York: Peter Lang. 2008.

Buckingham, D. *Beyond Technology: Children's Learning in the Age of Digital Culture*. Cambridge UK: Polity Press 2007.

Buell, L. *The Future of Environmental Criticism: Environmental Crisis and Literary Imagination*. Cambridge UK: Wiley Blackwell 2005.

Carvalho, A. 'Media(ted) Discourses and Climate Change: A Focus on Political Subjectivity and (Dis)engagement', *WIREs Climate Change*, 1(2): 172–179. 2010.

Clark, T. *Ecocriticism on the Edge: The Anthropocene as a Threshold Concept*. London: Bloomsbury. 2015.

Couldry, N. *Media, Society, World: Social Theory and Digital Media Practice*. Cambridge: Polity Press 2012.

De Jong, W., Shaw, M., and Stammers, N., *Global Activism, Global Media*. London: Pluto Press. 2005.

Dean, Jodi. *Post-Politics? No, Thanks! Future Non Stop*. Cambridge: Polity Press. 2011.

Deci, E. L., and Ryan, R. M. 'The "What" and "Why" of Goal Pursuits: Human Needs and the Self-determination of Behaviour' *Psychological Inquiry*, 11: 227–268. 2000.

Dunlap. R. E. 'The New Environmental Paradigm Scale: From Marginality to Worldwide use' *The Journal of Environmental Education*, 40(1): 3–18. 2008.

Feltham, O. *Alain Badiou: Live Theory*. London: Continuum. 2008.

Friemel, T. N. and Singer, S. 'Web 2.0 Literacy: Four Aspects of the Second Level Digital Divide' *Studies in Communication and Science*, 10(2): 147–170.

Garforth, Lisa Green *Utopias: Environmental Hope Before and After Nature*. Cambridge UK: Polity Press. 2018.

Garrard, G. *Ecocriticism: The New Critical Idiom* (second edition). London: Routledge. 2012.
Hall, S. 'The Determinations of News Photographs'. In S. Cohen and J. Young (eds), *The Manufacture of News: Social Problems, Deviance and the Mass Media,* rev. edn, vol. 4, Communication and society. London: Constable. 1981.
Hammond, P. *Climate Change and Post-Political Communication: Media, Emotion and Environmental Advocacy*. London: Earthscan Routledge. 2018.
Hobbs, R. *Digital and Media Literacy: Connecting Culture and Classroom*. California: Corwin Publishers (Sage). 2011.
Hulme, M. 'Cosmopolitan Climates' *Theory Culture & Society*, 27(2–3): 267–276. 2010.
Jenkins. H. *Convergence Culture: Where Old and New Media Collide*. New York: New York University Press. 2006.
Jenkins, H., Purushotma, R., Wrigel, M., Clinton, K. and Robison, A. J. *Confronting the Challenges of Participatory Culture: Media Education in the 21st Century*. Cambridge MA: MIT Press. 2009.
Kellner, D. *Media Culture. Cultural Studies, Identity, and Politics Between the Modern and the Postmodern*. London and New York: Routledge. 1995.
Koltay, T. 'The Media and the Literacies: Media Literacy Information Literacy, Digital Literacy' *Media Culture and Society* 33(2): 211–221. 2011.
Lapinski, M. K. and Rimal, R. N. 'An Explication of Social Norms' *Communication Theory*, 15: 127–147. 2005.
Lievrouw, L. and Livingstone, S. *Handbook of New Media: Social Shaping and Consequences of ICT's*. London: Sage. 2002.
Litzinger, R. 'Government from Below: The State, the Popular and the Illusion of Autonomy' *Positions: East Asia Cultural Critique*, 9(1): 206–253. 2001.
Livingstone, S. 'Setting the Scene for the Media Literacy Debate'. 2011. Available online at: www.gla.ac.uk/media/media_231178_en.pdf.
Livingstone, S. and Sefton-Green, J. *The Class: Living and Learning in the Digital Age*. New York: New York University Press. 2016.
Lunt, P. and Livingstone, S. *Media Regulation: Governance and the Interests of Citizens and Consumers*. London: Sage 2012.
Martens, H. 2010. 'Evaluating Media Literacy Education: Concepts, Theories, and Future Directions' *Journal of Media Literacy Education* 2: 1–22. 2010.
Methmann, C. 'Climate Protection as Empty Signifier' *Millennium: Journal of International Studies*, 39(2): 345–372. 2010.
Metzger, M. 'The Study of Media Effects in the Era of Internet Communications'. In R. Nabi and M. B. Oliver (eds) *Sage Handbook of Media Processes and Effects*. London: Sage. 2009.
Moore, E. *Landscape and the Environment in Hollywood Film: The Green Machine*. Basingstoke UK: Palgrave Macmillan. 2017.
Napoli, D. 'Revisiting "Mass Communications" and "Work" of the Audience in the New Media Environment' *Media Culture & Society*, 32(3): 505–516. 2010.
Nightingale, V. 'Television Audiences: Publics, Markets, Communities and Fans'. In John D. H. Downing (ed.) *Sage Handbook of Media Studies*, pp. 227–249. Thousand Oaks, CA: Sage. 2004.
Nisbet, M., Ho, S., Markowitz, E., 'Neill, S., Scihafer, M. S. and Thaker, J. (eds). *Oxford Encyclopaedia of Climate Change Communication*. Oxford: Oxford University Press. 2018.
Peer, S., Goldnam, D. and Yavetz, B. 'Environmental Literacy in Teacher Training: Attitudes, Knowledge, and Environmental Behaviour of Beginning Students' *Research and Reports*, 39(1): 45–59. 2007.

Peter, J. and Valkenburg, P. 'The Effects of Internet Communication on Adolescents' Psychosocial Development: An Assessment of Risks and Opportunities'. In E. Scharrer (ed.) *Media Effects/Media Psychology* (*The International Encyclopedia Of Media Studies Vol. 5*), pp. 678–697. Hoboken, NJ: Wiley-Blackwell. 2013.

Potter, J. 'The State of Media Literacy' *Journal of Broadcasting and Electronic Media*, 54(4): 675–696. 2010.

Rhengold, H. *The Virtual Community: Finding Connection in a Computerised World*. Reading, MA: Addison-Wesley. 1993.

Ritzer, G., Dean, P. and Jurgenson, N. 'The Coming of Age of the Prosumer American'. *Behavioural Scientist*, 56(4): 379–398. 2012.

Rosen, J. 'The People Formerly Known as the Audience' PressThink: Ghost of Democracy in the Media Machine. 27 June 2006.

Roth, C. E. *Environmental Literacy: Its Roots, Evolution and Directions in the 1990s*. ERIC – Clearing house for Science, Maths and Environmental Education. Columbus: Ohio 1992.

Roth, C. E. 'On the Road to Conservation'. In *Massachusetts Audubon* pp. 38–41. 1968.

Rust, S., Monani, S., and Cubitt, S. (eds) *Ecocinema Theory and Practice*. London: Routledge. 2013.

Silverblatt, A., Ferry, J., and Finan, B. (eds) *Approaches to Media Literacy: A Handbook*, 2nd edn. New York, London: M.E. Sharpe. 2009.

Swyngedouw, E. 'The Non-Political Politics of Climate Change' *ACME: An International E-Journal for Critical Geographies*, 12(1): 1–8. 2013.

Urry, J. *Climate Change and Society*. Cambridge: Polity 2011.

Vandoninck, S., d'Haenens, L. and Roe, K. 'Online Risks: Coping Strategies of Less Resilient Children and Teenagers across Europe' *Journal of Children and Media*, 7(1): 60–78. 2013.

Vansteenkiste, M. and Ryan, R. M. 'On Psychological Growth and Vulnerability: Basic Psychological Need Satisfaction and Need Frustration as a Unifying Principle' *Journal of Psychotherapy Integration*, 23: 263–280. 2013.

Webster, J. G. 'The Audience' *Journal of Broadcasting & Electronic Media*, 42(2): 190–207. 1998.

2 Understanding audience psychology and trigger points for promoting environmental literacy

Overview

This chapter draws on a broad range of psychological and behavioural studies which have been used to explain the lack of take-up of environmental agendas culminating in the spectre of climate change catastrophe across the world. Understanding audience psychology and behaviour trigger points remain an essential prerequisite towards developing strategies for effective environmental literacy programmes. In particular, drawing on the critical literature developed across the broad-church of eco-literature – which can be characterised as oscillating between so-called first and second wave environmentalism – helps to tease out modes of best practice that promote new forms of environmental literacy. This environmental critical process will be illustrated through an analysis of the opening episode of a successful 2017 Netflix series titled *Ozark*.

Psychological tools for understanding audience reactions

'It is increasingly clear that understanding human responses to climate change is just as important as – if not more important than – understanding climate change itself' (Weintrobe 2013: 1). Most people generally accept anthropogenic global warming yet continue to locate it as a problem of the future and actively try to negate it as being an urgent problem, not needing immediate attention. The big question for environmental media scholars across so many existing new platforms remains, how can mass media serve to signal and even promote new forms of low carbon energy futures and hopefully encourage radically transformative modes of environmentally driven behaviour?

Recalling for instance debates around the dangers of cognitive dissonance,[1] climate psychologists continually affirm that audiences may think something intellectually, while not being emotionally connected to pressing issues around climate change in particular, and ostensibly not seeing this as something to do with them personally. Scholars posit at least three forms of denial, which ought to be taken into account and addressed when examining mediated stimuli's role in sparking audience behaviour and at the same time concentrating on engaging with corrective levels of environmental literacy and effective human agency:

> Denialism – campaign of misinformation regarding climate change
> Negation – which is reminiscent of the first stage of mourning; this involves denying something that is by all accounts true, or
> Disavowal – knowing and not knowing at the same time.
>
> (Weintrobe 2013: 7)

Weintrobe goes on to clarify what she characterises as 'survival anxieties' that also weigh down audience thinking including:

> 1 'depressive anxieties' and the way we 'face the loss of the Earth as a dependable bedrock that enables and supports our very life, the life of other fellow species and future life for all. Specifically, we face the effects of a climate tipped into instability'.
> 2 'When we register that the climate is out of kilter and are faced with what Friedman (2008) calls "climate weirding", at stake is not only anxiety about our physical survival, but the survival of our very sense of self.'
> 3 Furthermore, there is an over-dependence on leaders and politicians to 'solve things'. Many commentators speak of the abiding iconic image of the failed Copenhagen Climate Conference and the perceptions 'of truth bending the knee to greater, more hidden sources of power in the wings'. While the relative success of the more recent Paris COP agreement may have swung the pendulum in a more productive direction, there still remain many political and other roadblocks to progress.
>
> (Weintrobe 2013: 42–43)

Despite being the most important societal challenges of the twenty-first century – (see for instance the caustic science fiction speech by Govern Nix in *Tomorrowland*[2]) – public engagement in the media around climate change remains surprisingly low, especially in the United States, while being more favourable regarded across several EU countries and other regions around the world. Mounting evidence from across the behavioural sciences has found that most people regard climate change as a non-urgent and psychologically distant risk, which in turn has led to deferred public decision making about mitigation and adaptation responses (see van der Linden *et al.* 2015: 758). Most certainly the psychological barriers Weintrobe and other behavioural experts talk about, are having a deleterious effect on climate change communication and the prospect of developing corrective behaviour patterns.

We must at least consider the possibility that the public remains disillusioned and paralysed by emotionally-loaded doom type scenarios, including those apocalyptic fictional films around climate change. As shown in research across other dystopic catastrophes, publics often appear to become indifferent and turn away, displaying signs of compassion fatigue among other symptoms, from what are considered incomprehensible events. Rational engagement with justice issues for example and doing the right thing for the future survival of the planet have to be married with the essentialising pull and understandable psychological response of emotional disengagement and an overall lack of connectivity. Yet as Hoijer (2010) most notably states: 'social engagement is based on the capacity to feel', which is a prerequisite for the exercise of reason. At the same time, we must always keep in mind that 'reason without emotion does not lead to action' (2010: 728).

As frequently illustrated in the communications literature, promoting various forms of emotional engagement and environmental connectivity remains one of the core attractions and strengths of audio-visual narratives (see Brereton 2005, 2016; Weik von Mossner 2017). Developing emotional, as well as scientific literacy and empathy, while encouraging clear identification with such troublesome topics, often beyond our immediate reach, remains a core strength of such appeal and helps demonstrate the unique affordances of literary and especially audio-visual narratives.

At the same time, psychological research has consistently shown how negative effects can also be one of the strongest drivers of climate change risk perceptions and policy support (Leiserwitz 2004). In short, how we feel about a given situation often has a potent influence on our decisions about how to respond (Slovic and Peters 2006). Statistical descriptions of the risk of climate change however often fail to elicit concrete action, because statistical information by itself, means very little to (most) people. Alternatively, vicarious emotional and digital sensory experience through the medium of powerful narratives, even without the luxury of first-hand experience, can become a much more engaging, albeit surrogate teacher.

For example, although the odds of death or injury from a terrorist attack in the United States are very low, terrorism is still ranked as a top (inter)national priority, whereas the threat and reality of climate change is not. The difference probably lies in the fact that for terrorism, vivid, memorable and emotional experiences readily come to mind, recalling nationalistic fervour and patriotism. Most notably this response is evident through the spectre of 9/11 and its aftermath in America, which remains in the public consciousness, especially with ongoing concerns over global terrorist groups like ISIS. Concurrently, the role of unconscious visceral experience – as dramatised in televisual series like *Homeland* to be discussed in a subsequent chapter – catalyses powerful emotional engagement and connectivity, while speaking to political and even environmental concerns. Unfortunately, this effect has largely been ignored within climate policymaking, resulting in further alienation for citizens. Closely linking climate change to various other political and social issues is important to keep green issues firmly in the public consciousness. Popular media story-tellers can assist in this endeavour, especially taking into account the difficulty that this phenomenon is generally considered as a slow-moving, 'invisible' process that cannot easily be experienced directly. This book is founded on the belief that it is difficult to achieve media traction without the assistance of fictional narratives. Finding effective visual correlatives and narrative tropes that both engage audiences, especially those not predisposed to climate change debate, while also educating without overtly preaching to all citizens across the world, has to become a key element of communications and a defining marker towards promoting environmental literacy.

Furthermore, because climate change is a global problem with global consequences and being very difficult to solve, peoples' sense of *personal efficacy* (i.e. the belief that individual actions can make a difference) unfortunately remains relatively low (Kerr and Kaufman-Gilliland 1997). Indeed, the global nature and complexity of the climate change problem tends to make people feel powerless and hence more immersive and personalised engaging narratives are essential to help kickstart and bring such issues onto the centre stage. By all accounts, it is however often more effective to appeal to and leverage the social context within which citizens make decisions, particularly in helping promote *collective efficacy* – recalling the belief that group actions can make a difference (Roser-Renouf *et al.* 2014).

Much of the media, and even the scientific and policy discourse around climate change, has consistently invoked the idea of losses. For instance, climate solutions are often framed as an immediate loss for society (e.g. higher taxes, reducing energy consumption and resultant poorer standards of living). Yet, long-standing behavioural research has shown that humans psychologically evaluate gains and losses in fundamentally different ways. This is brought into focus for example since climate change impacts are often framed as somewhat uncertain, with losses in the distant future, whereas climate change solutions are seen as constituting certain economic losses for society at present. Consequently, framing such a balance sheet logic encourages some people to conclude that maintaining the status quo might almost be 'worth the gamble' in the longer term.

Those with power and wealth can always for instance move inland or upland to escape flooding or avoid some other dire consequences of climate change – as witnessed in various scenarios embedded within several science fictional narratives. See for instance the somewhat risible science fiction film *2012* (2009). Hence such clearly articulated long-term, but apparently slowly evolving consequence of climate change are not easily faced up to, much less addressed, through consensual public support. Consequently, if only unconsciously adopting such logic, citizens can simply ignore the problem.[3] Meanwhile, the timescales for orchestrating democratic politics, which need to secure consensus to build efficient and sustainable behavioural change, appears to be far too slow, remaining constantly hampered by the short-term cycles of electoral governance. Nonetheless, effective emotionally-laden narratives can greatly help push beyond these barriers, characterised by a utilitarian causal-effect and rationally-privileged mindset that pulls against taking on board the urgent policy and behavioural change required to face up to such catastrophic eventualities.

At the same time, communication psychologists further distinguish between two separate sources of motivation: extrinsic and intrinsic, which in turn can be mapped across varying coping and therapeutic strategies (Stoll-Kleemann *et al.* 2001), or more emotional management strategies (Hochschild 1979), all the while further recalling the fear of hopelessness or apathy (Lertzman 2008). Media narratives appeal to all of these often-contradictory feelings and emotions in varying measures. Meanwhile, most citizens no matter what their psychological make-up or ideological predisposition across the Left–Right spectrum, continue to have at their disposal an extensive array of psychological strategies to help them suppress thoughts and feelings about anxiety-producing situations, including a pervasive sense of threat, emanating from the multimodal implications of the effects of climate change.

While somewhat counter-intuitively however, Lertzman (2008) has commented that

> what can appear as public apathy about environmental issues is the result of people caring too much, not too little about what is happening. We are poorly equipped because of our very nature as human beings to bear the truth that global warming is mostly caused by human activity. This is because the truth makes us anxious.

The truth about damage to climate stability also makes many humans feel guilty and ashamed, and whether or not it is rational to have these feelings, having them is also part of human nature. 'Our main reason we find guilt and shame such difficult emotions to bear, is that they cause us considerable anxiety' (see Weintrobe 2013: 45).

Furthermore, at another level of specific environmental exploration and analysis, Mike Hulme – a very influential scholar in the field – remains most sceptical of the strategy of constructing climate change as the 'mother of all problems' and the '[greatest/defining/most serious] long-term [problem/challenge/treat] facing humanity' (2009: 333). Many studies of environmental communication and climate change, including this one, can be equally accused of aggregating too many issues under the broad rubric of environmentalism and climate change, while promoting over-expectations around the potentiality of media coverage to address such concerns. While such cautioning has certain validity, nonetheless, if one accepts the foundational principle of 'everything being connected' that underpins much deep ecological thinking, then it becomes more difficult to tease out clear boundaries and demarcations, as Hulme highlights. Perhaps, contemporary society has out-manoeuvred itself, by embracing the overall inability of securing effective political action to address this apocalyptic situation. Hulme appears right, however, in suggesting

> we have allowed climate change to accrete to itself more and more individual problems in our world – unsustainable energy, endemic poverty, climatic hazards, food security, structural adjustment, hyper-consumption, tropical deforestation, biodiversity loss – and woven them together using the meta-narrative of climate change.
>
> (Hulme 2009: 333)

Most certainly this remains an ever-present danger when trying to develop a robust, yet discreet arsenal within an environmental literacy tool box to incisively address the nexus of clearly defined environmental issues encapsulated by climate change, much less in trying to tease out a clear hierarchy of priorities across so many environmental issues. Unfortunately, there is no magic bullet formula and different audiences in varying regions of the world experience different levels of climate problems, while coincidentally being exposed to a varying diet of environmental media consumption. Consequently, all pedagogical strategies have to be contingent with and measured across a wide range of cultural and other contextual variables.

Adopting a broad-brush approach and an ethical philosophy, encapsulated by the catchphrase 'everything is connected' makes this educational process even more difficult to tie down, affording no easy media solutions. Nonetheless, categorising and de-marking a range of related issues around climate change, needs to be constantly monitored and evaluated to help ensure a clear sense of scale and structured division across the full gamut of environmental concerns. Not to mention teasing out and closely appreciating the evolving psychology of audiences and behaviour patterns that map onto general media consumption. Within communications studies there is an evolving normalising process of assessing the power of media to influence and affect a whole host of audiences, which is often characterised as effects theory.

Effects theory, technological determinism and environmental literacy

Historically, some educationalists suggest the primary purpose of media literacy is to enable people to defend and inoculate themselves against the 'big bad propagandistic media', by affording them critical skills, so they can recognise and diffuse the power of persuasive, manipulative and generally exploitative media. Alternatively, as also alluded to in the introductory chapter, more recently the pendulum has swung towards empowering publics and students to participate in a multi-mediated world, providing them with the analytic and creative resources they require to benefit from the wealth of ideas and inventiveness that media can offer. From this perspective, the more developed the tools of media literacy there are, the better the media use which can result as a consequence, while at the same time, aiming to promote 'more diverse, thoughtful and engaged' citizens (Jenkins 2006).

Literacy at every level and sphere of influence does not simply demand knowledge and the ability to encode and decode messages through various media, but rather demands 'applying this knowledge for specific purposes in specific contexts and use' (Schriber and Cole 1981: 236). Media reception and effects debates have become particularly pointed and developed with the expediential growth of new and more interactive media formats. Consequently, the unique affordances of various media and technological literacy, necessary to become proficient in applying new technological and other skills, has reached a very high level of expectation and competency.[4]

For instance, online data analytics of users have become a very important tool for new media industries to measure, assess and evaluate their overall success. By all accounts their long-term sustainability is linked to monetising and harvesting user profiles for advertising or political purposes.[5] Such skills and affordances are most especially built into online surfing, such as evident on the YouTube platform and other formats like video games, explored across a number of chapters. While earlier chapters will focus on the apparently less sophisticated threshold of technological and interactive affordances required by more conventional media, such as documentaries, fictional film and online streaming television, later chapters will concentrate on video games, alongside YouTube and Netflix streaming of niche media texts.

Appreciating media audience behaviour: the uses and gratification model (U&G)

The uses and gratification model is frequently referenced as the most pertinent framework for understanding contemporary media selection and usage. The model has evolved to include a number of primary tenets, which remain overly schematic and static. Nonetheless, individuals appear aware of their social, psychological, and biological needs, 'to evaluate various media channels and content; assess functional alternatives; and to select the media or interpersonal channel that they believe will provide the gratifications they seek to meet their various needs' (see Blumler and Katz 1974). Such general needs include those related to diversion (e.g. escapism, arousal), personal relationships (e.g. social utility), personal identity (e.g. reality exploration), and surveillance (e.g. news gathering), alongside one could add, political and environmental sustainability – all of which can be explored through a broad range of fictional media outputs examined throughout this volume.

The U&G perspective which underpins many of the readings and subsequent chapters remains grounded in the conceptualisation of audience members as active, in control of their own media consumption and able to provide accurate self-reports about the gratifications they seek and receive from the media. Such a model would however appear to pull against much of the psychological literature discussed above, around general apathy with regards to climate change and taking into account resistance towards facilitating real behavioural change. Critiques of the model include assertions that it is overly-individualistic in focus, lacks coherent typologies of people's motives for media use and how it probably needs greater clarity in teasing out some of its central concepts. Additionally, some scholars have questioned the assumption that individuals are necessarily even aware of their inherent viewing motivations (Zillmann 1985, cited in Nabi and Krcmar 2016: 2).

Furthermore, Gerbner's extrapolation of cultivation theory draws on related beliefs that the primary function of media is to stabilise rather than to challenge behaviour patterns, whereas most social science research aims at measuring the alterations in beliefs and behaviours induced by exposure to media messages.

The theory of cultivation helps to describe the cultural function of media as a shaper of social reality. According to such scholars, the primary impact of the media therefore serves 'to reinforce, not to challenge the structure of power', while the 'function of storytelling in any society is to fit reality to the social order' (Morgan *et al.* 2012: xii). Some more recent scholarship does not slavishly accept this premise and remains more positive with regards to the transformative power of media.

Meanwhile, research on parasocial relationships (PSRs), for instance, suggests that people primarily use media to fulfil their desire for social connections to others. Such appeals around interconnectivity can be extended however – on a metaphorical level at least – to include engaging with deep ecological values that underpin and promote the 'interconnectivity of all living things'. This pervasive sense of global connectivity is illustrated across several chapters in this volume. Such tensions are encapsulated by recalling the historical and symbolic power of *Casablanca* (1942) in highlighting tensions between (isolationist) individuals, as opposed to communal and group solidarity, while the world historically faced global warfare, up to the more localised and aesthetically small-scale niche example of Prince Ea's rap YouTube video *Sorry*, examined in Chapter 10. Consequently, many examples and illustrations of environmental issues and concerns, call for more global interconnections and especially towards recognising the need for inter-generational justice across such global problems. Over the years a number of engaging narratives have been produced for mass audiences across all media formats and platforms. These can be fruitfully re-read, at least against the grain, as speaking to such broadly-based environmental values and at least calling out the possibility of promoting radical environmental models of behaviour.

Lull (1980) for example offers a comprehensive discussion of the social uses of television and the family (which can be extended to other forms of media), noting that it serves two categories of functions: structural and relational. Structural functions include environmental (e.g. background noise, companionship) and regulative (e.g. punctuating time, reward). Furthermore, within relational functioning, Lull identifies four categories: communication facilitation, affiliation/avoidance, social learning, and competence/dominance (Lull 1980: 3). Much of these wide-ranging attributes resonate across a number of chapters in this volume, including influential televisual series from *The Sopranos* to *Better Call Saul*, which are explored in subsequent chapters.

Yet as behavioural and psychological studies infer, by helping to improve public decision-making about climate change, audiences and citizens should reject how the phenomenon has traditionally been reductively framed – primarily as an analytical, temporally and spatially distant risk that represents an (uncertain) future loss for society. Much psychological research alternatively suggest that in order to improve public engagement with the issue, policymakers should instinctively emphasise climate change as representing 'an experiential, local and present risk; define and leverage relevant social group norms; highlighting the tangible gains associated with immediate action; and last, but certainly not least, appealing to long-term motivators of pro-environmental behaviour and decision making' (van der Linden *et al.* 2015: 763; Devine-Wright 2004; Klas *et al.* 2018). All of these representational strategies are very far from the more tentative and ambiguous messages of much conventional media output across the public sphere.

As memorably articulated by Timothy Morton in *Being Ecological* (2018), with regards to growing worries over plastics in the oceans, which has recently captured the public imagination, we need to 'sense the squiggling in your gut. Now visualise all the plastic fragments in all the oceans of the world. Can you see them, feel them? They are not "out there", they are interconnected with us in time and space' (77). Audio visual narratives and new fictional imaginaries across so many media formats can in turn creatively help audiences to actively engage with a wide range of evocative environmental issues and lessons. Making such environmental issues real, personal and locally relevant, is what much eco-media scholarship strives to promote.

Mass media permit such connections to also yield to beauty and disgust, to an acceptance of enchantment and mystery as being essential aspects of long-established forms of rationality. For Morton, ecological ethics is

> about expanding, modifying and developing new forms of pleasure, not restraining the meagre pleasures we already experience, because we are only thinking in ways that our current modes of doing things allows. What would pleasure look like beyond the oil economy for example.
>
> (2018: 152)

Such a trajectory must also surpass scientists' approach to activism and empirical adaptation of 'data dumps', helping to prove the severity of climate change, but often failing to engage with the population at large. Most certainly the media-sphere requires a big metaphorical and representational jump of its imaginative potential for audio-visual narratives to help uncover their deep environmental voice.

To historically illustrate some of the psychological and other concerns raised above with regards to audience effects, while beginning to tease out new strategies for environmental literacy and education, the following section will outline an evolving ecocritical historical debate. This is encapsulated by first and second wave environmental analysis, which in turn illustrates an eco-textual reading of a recent Netflix series *Ozark* that concludes this chapter.

First versus second wave ecocriticism and audience effects

Lawrence Buell highlighted the well-honed distinction between first and second wave ecocritical approaches in his 2005 study, *The Future of Environmental Criticism*. Buell distinguished between older (generally speaking, twentieth-century) environmental criticism, which was often preoccupied with nature writing, wilderness and texts such as Thoreau's *Walden*. As opposed to a broad range of twenty-first century work that is generally more concerned with a variety of landscapes (including places like cities) and more contemporary environmental issues, encapsulated by the spectre of climate change.

Essentially, a second-wave environmental approach and criticism is careful not to overly romanticise wilderness (as did some of their predecessors), while being more likely to direct themselves to sites of environmental devastation and texts that do the same, such as Rachel Carson's *Silent Spring*.[6] Consequently, romantic writers such as Thoreau and Wordsworth, (and the audio-visual contemporary equivalent of romantic nature and environmental documentaries, such as the David Attenborough's BBC television franchise) who are the darlings of first-wave environmental criticism, have generally received less attention from second wave criticism.

However, from a contemporary eco-film and general media literacy perspective, this somewhat crude binary division remains unhelpful. New-generational media audiences (and even scholars) in the process of becoming literate and engaged within an environmental perspective, probably need to go through the initial romantic and biophilic stages of nature celebration, which in turn corresponds with so much mainstream generic evocations that dramatise the majesty of nature and landscape appreciation. The more advanced critical and reflexive stage of environmental engagement I suggest is less evident at least within mainstream mass media culture. As one might expect this is more specifically evident within experimental and *avant garde* output, but unfortunately this garners relatively small audiences.

Recalling for instance Rachel Carson – especially in the pastoral opening of *Silent Spring* – where she intentionally romanticised nature as a rhetorical strategy, designed to enlist readers to combat threats to the environment. This rhetorical approach appears very different to so-called first-wave environmental criticism. Alternatively, a second-wave eco-critic would generally not be led by Carson into making a fetish of nature (as sometimes happened in the first wave). Such scholarship uncovers how such romanticising takes place as a representational construct. This direct critical approach calls attention to the manner in which this rhetorical strategy influenced the first wave of environmental critics, who were in some cases blind to its overall influence (Hiltner 2015: 131).[7]

One could almost assert that had an interest in ecocritical approaches not been actively fostered in the closing decades of the twentieth century – which in practical terms, resulted in academic courses and other programming across colleges and universities – second wave ecocriticism would not have had acquired a firm foundation on which to build its radical and more critical discursive practice. This trajectory, albeit on a less dramatic scale, can also be recognised when analysing the evolution of media and film literacy studies. Nevertheless, across media platforms, so-called first wave environmentalism, espousing an unadulterated love of nature, continues to have as much appeal and influence through what can be characterised as a primal love and deep connection with (wild and exotic) nature. However, the nature documentary format like that exemplified in *The Blue Planet*, for instance, tangentially at least remains coupled with a more reflexive agenda around environmental conservation and most notably waste for example. This has been recently highlighted by the scourge of plastics posing a major new problem for the survival of our oceans and planet. Both first and second wave modes of environmental engagement are consequently not so easily separated and by all accounts have become even more intimately cross-connected. This sense of continuity rather than dissonance across environmental expression and learning is illustrated by a popular recent Netflix series titled *Ozark*, which in varying ways play off a range of signifiers and tropes that feed into both first and second wave modes of expressing an environmental sensibility.

Ozark: going back to nature using a contemporary critique

The Netflix series created by Bill Dubuque certainly has clear echoes of *Breaking Bad* to be discussed in a subsequent chapter. Audiences are introduced to what seems like a boring city-based financial consultant Marty Byrde (Jason Bateman) and his wife Wendy (Laura Linney), together with their 'normal' teenage family. While trying to sell his services to a young couple who apparently know nothing about finance, Byrde gets an email and an attachment, which reveals a woman having passionate oral sex. One wonders at first if the character is into cheap porn, as also noticed by his business partner Bruce Liddell (Josh Randall), who later confronts him with this very question. The two partners who are considering moving to bigger premises, are shown a fantastic all-glass, high-rise office block in Chicago, with a magnificent view circumventing the city. Byrde is more tentative and cautious in coming to a decision, and probably remains more concerned with environmental sustainability than his partner, remarking on the cost of heating the offices and noting that they are on the 'dark side' of the building.

Meanwhile, his more flamboyant partner Liddell appears bemused at such concerns and out of the blue presents Byrde with a tourist brochure for a place called Ozark. Like the earlier 'porn video', this coded signifier and piece of narrative exposition has transformative significance, as evident later in this enticing opening episode. All of these narrative cues (recalling David Bordwell *et al.* 1985 classic exposition of the workings of a Hollywood narrative and how they function for audiences) demonstrate the creative and innovative power of new forms of intertextual elements and their complex modes of development, using what might be considered a mash-up of smart narrative tropes. The tattered paper brochure is reminiscent of the 'paradise falls' motif in Pixar's classic *UP* (2009) – a utopian environmental place where one should aim to visit and even escape to before one dies (see review in Brereton 2012: 141). But in this instance, it is revealed as a real-estate space and commercial opportunity for success, far away from the corruption, regulation and control of money centred in this growing metropolis of Chicago.[8]

This spoiler alert hook of escape to a predefined (ecological) nirvana – which underpins so much first-wave biophilia from *Walden* to *Star Wars* – recalling of late the mythical island beauty of Skellig Michael in Ireland, highlighted in the latest episodes of this global franchise – is further developed in the most over-the-top and violent reveal scene in any television streaming/series. Here the partner's nefarious criminal business activities are suddenly explained.[9] Under extreme torture, his partner Liddell admits he was skimming over eight million dollars from the money-laundering business set up by an evil drug lord boss, played by Esai Morales. Shockingly, in an instant, everyone is shot dead for such double-crossing, except the main protagonist, who under severe psychological and mortal pressure 'weaves a story' that he will return the money and earn much more besides, if let move home to this 'magical Ozark forest place'. This is set up as a utopian place where all the rich come to party in the summer and has so much access to a precious nurturing and fertile lake, surrounded by an over-hanging and protective forest. A (fantasy) tale almost re-spoken word for word by Byrde, while under a death threat, recalling the now dead partner's somewhat risible speech a few scenes earlier.

But what has all this narrative exposition got to do with environmentalism? The final shots of this stunning opening episode at last visualise this environmentally pristine nature space/place being talked about. Having driven so long and literally having his world taken from underneath him, Byrde is so traumatised, he gets out of the car to relieve himself in the woods. Walking on a bit through an opening in the forest, he initially spies the magical and mystical lake, way down underneath him through the forest clearing. The fractured and traumatised family join him – reminiscent of the final closure in an eco-reading of *Grand Canyon* (1991) (see Brereton 2005) – the camera self-consciously pulls out in dramatic and predefined stages, punctuated by orchestral music and a memorable soundscape, to frame their look and reverie, as the protagonists witness the sublime vision of an unspoilt landscape with a massive lake system underneath them, together with a rich verdant forest all around.

It's as if they have come out of the primeval forest for the first time – reminiscent of the journey visualised in *Apocalypto* (2006), alongside recalling so many eco-nature or Western films – and discovered a vision of a potentially bright future, where 'raw nature' and a benevolent unspoilt habitat might be their final saviour, as they strive to psychologically repair themselves from the awful trauma that has befallen them. But unlike the closed narrative elements of conventional fictional films, this is just the opening teaser episode in a series which spends hours explaining the physical attributes and implications of their (Walden-like) journey 'back to nature' and entry into this magical place. The storyline consciously strives to critically oscillate between first and second wave modes of engagement and even, I would argue, insinuating various modes of eco-critical learning from its active engagement with nature.

The American (first-wave environmental) dream traditionally was built on going out 'West' to discover a pristine natural setting to help 'grow a family'. Such myths and perennial tropes and narrative trajectories continue with a critical inflection embedded within such new media formats, adding an innovative spin to the cosmopolitan-versus-rural binary opposition. One might even suspect the narrative invites renewed first-hand interest and broad appeal for many new generations of audiences. This consequently speaks to a deep environmentalist constituency, as many disaffected city-folk, metaphorically at least, desire to escape to the (idealised) countryside and adapt a simpler and hopefully more sustainable style of life. This transformation can include adopting to the demands of a low carbon energy future. But recalling the tragic and sometimes surreal results of classic escape to the countryside within film history – from the conventional horrors of Kubrick's *The Shinning* (1980), to *avant garde* exploits in the lakeside country retreat in Von Trier's fraught *Funny Games* (2007) (see Brereton 2016: 104/5) – many filmic narratives like this Netflix televisual series actively illustrate the horrific reality, as well as the fictional fantasy of such simple-minded one-directional environmental escape.

At the same time, such storylines continue to provoke a fresh appeal for the latent ecological meaning of such habitats for new generational audiences, adding to the pervasive hope and appeal of 'going into nature', while recalling the inherent power of environmental connectedness. This well-worn trajectory is celebrated across the visual arts and especially within the seminal poetry of Wordsworth or Coleridge for example, alongside the nature writings of Thoreau and other new forms of (cli-fi) literature. On reflection, one wonders in future episodes how an 'untouched' pristine touristic image, such as the lake of the Ozarks in Missouri, can be so effectively visualised and potentially re-colonised from an environmental perspective? This audio-visual narrative at least sets up such oscillating tension, while moving beyond simple first wave romantic celebration of nature and at least unconsciously calling to mind new modes of exploring nature and related environmental concerns which deserves dedicated textual analysis.

Concluding remarks

Public awareness of humanity as the dominant agent of instigating global change, much less recognising people's primary role in causing climate change, always remains in tension. Preparing for such eventualities is now unavoidable, as global society tries at least to imagine moving to a low carbon energy future. Consequently, the role of environmental learning and developing new forms of literacy, framed within cautionary and engaging audio-visual narratives, has a major role to play in this process.

Citizens need to understand and appreciate the implications of the truly daunting global environmental challenges facing us all. But they also require evidence and creative imaginative examples and narrative illustrations that can be used to help uncover a fruitful path towards an enduring form of (post-capitalist) prosperity. At the same time, we need to celebrate, promote and publicise progress along this journey, while never forgetting that humanity has embarked on a different relationship with earth, its only home (Dalbotten *et al.* 2009: 7). Media and communication models of engagement and reception have a major role to play in helping all societies realise the need for such necessary transformation and in helping to tangibly comprehend how such dramatic paradigm shifts in our ways of life might be realised. Metaphors, creative imaginaries and other strategies used in audio-visual media and narratives, including the more conventional appeal of biophilic escape evident in *Ozark*, serve as a key element of this process. Audiences continuously crave more creative images and eco-narratives of a post-carbon future, together with more positive references, towards exploring new ways of engaging with technology, while appreciating and understanding our ever-changing environment and habitat.

Finally, whether categorised as audiences, citizens or stakeholders, we all need to recognise that the problem of human influence on the planet is about more than simply addressing climate problems to help humanity survive, it is about how our varying behaviour affects all aspects of life (Kiehl 2016: 9). Most notably Fritzof Capra has for over 30 years been telling us how the world is interconnected, and how we cannot simply use the economic measure of GDP to inform us of the true happiness of the person or the planet.

> Just as biodiversity is an essential component of ecological sustainability, so is cultural diversity essential to social sustainability. Diverse values should not be respected just because we are tolerant folk, but because we must have a pool of diverse perspectives in order to survive, to adapt to changing conditions, to embrace the future.
>
> (Hawkes 2001: 14)

Embracing the fluid boundaries between first and second wave and future developments in environmental study, coupled with appreciating and learning from psychological tools to measure effective engagement with audiences are all badly needed as we imagine the future media landscape. The following chapter will interrogate related aspects of this jigsaw while taking on board the development of new forms of environmental literacy and how greenwashed imagery might alternatively be repurposed in more productive ways.

Notes

1 Cognitive dissonance, for example (see Festinger 1957), may arise when a parent engages in and holds inconsistent behaviours, beliefs, values and ideas. As a result, parents who allow their children to consume media must derive a valid set of beliefs and motives to avoid dissonance. Consequently, beliefs may not directly drive motives (Nabi and Krcmar 2016: 15).

2 Governer Nix's (Hugh Laurie's) speech in *Tomorrowland*:

> Let's imagine … if you glimpsed the future, you were frightened by what you saw, what would you do with that information? You would go to … the politicians, captains of industry? And how would you convince them? Data? Facts? Good luck! The only facts they won't challenge are the ones that keep the wheels greased and the dollars rolling in. But what if … what if there was a way of skipping the middle man and putting the critical news directly into everyone's head? The probability of wide-spread annihilation kept going up. The only way to stop it was to show it. To scare people straight. Because, what reasonable human being

wouldn't be galvanized by the potential destruction of everything they've ever known or loved? To save civilization, I would show its collapse. But, how do you think this vision was received? How do you think people responded to the prospect of imminent doom? They gobbled it up like a chocolate eclair! They didn't fear their demise, they re-packaged it. It could be enjoyed as video-games, as TV shows, books, movies, the entire world wholeheartedly embraced the apocalypse and sprinted towards it with gleeful abandon. Meanwhile, your Earth was crumbling all around you. You've got simultaneous epidemics of obesity and starvation. Explain that one! Bees and butterflies start to disappear, the glaciers melt, algae blooms. All around you the coal mine canaries are dropping dead and you won't take the hint! In every moment there's the possibility of a better future, but you people won't believe it. And because you won't believe it you won't do what is necessary to make it a reality. So, you dwell on this terrible future. You resign yourselves to it for one reason, because *that* future does not ask anything of you today. So yes, we saw the iceberg and warned the Titanic. But you all just steered for it anyway, full steam ahead. Why? Because you want to sink! You gave up! That's not the monitor's fault. That's yours.

3 For instance, his rationale is similarly applied to a lack of public buy-into taking on a personal pension, or for that matter any other form of insurance provision for the future. It follows the logic of why lose out today for something that possibly may only happen in the distant future and might even be alleviated in large measure over time or through some future technological innovation to clean CO_2 from the atmosphere. Yet a majority of citizens have accepted the necessity of having house fire insurance, while this rarely if ever happens. Why not take out environmental insurance, which is much more likely! Dealing with the multiple effects of climate change certainly requires the build-up of insurance buffers and security banks of scarce resources to help deal with a broad range of planetary problems, well into the future.

4 Just picking a random example from online media technology, users might authenticate themselves on the Internet Movie Database (IMDb), using Google and Facebook identifiers. If users choose to authenticate themselves with Facebook, they are redirected to the Facebook login page. Here the visual hierarchy of objects on the interface steers users towards the 'Allow' button, thus enabling IMDb to access all their private data and post on their behalf. Facebook and Google protocols similarly illustrate what users and audiences may call the illusion of a transparent interfaces, while the user's goal is simple, but the (technical) process of usage and the technical affordances of the software embeds the power of commercial interests.

5 This has recently led to global problems with the harvesting of digital information from Facebook for political ends, as illustrated by the 2018 'Cambridge Analytica' controversy.

6 Of course many environmental scholars talk of a third wave across literature and other modes of discourse. For instance, Scott Slovic apparently considers the second wave – which he dates to around 1995 – involved the broadening of the corpus of narratives studied and even went beyond literature itself, as the label 'green cultural studies' suggests (Slovic 2010). The third wave is envisaged as beginning in the early twenty-first century, as named by himself and Joni Adamson in a 2009 article in MELUS – see a summary essay on contemporary ecofeminism by Estevez-Saa and Lorenzo-Modia (2018: 136).

7 Incidentally, similar pedagogical and evolutionary strategies can be mapped by the long-term development of media literacy, which has been brought into the mainstream through formal education strategies and are supported by ongoing curricular innovation.

8 This trope also has echoes in numerous eco-critiques of big business and finance fables including *The Wolf of Wall Street* (see review in Brereton 2016), where the notion of money – like in *Breaking Bad* – is initially at least set up as the panacea towards securing

long-term happiness. This pervasive trope of material well-being is further reinforced through the futuristic opening fantasy credits of the series, with drugs being smuggled by Byrde using the disguise of fish and ice on a small boat in this erstwhile touristic nature setting.

9 Only much later in the series is this elicit activity and decision to work with gangsters teased out in a flashback episode.

References

Blumler, J. G. and Katz, E. *The Uses of Mass Communications: Current Perspectives on Gratification Research.* Beverly Hills, CA: Sage 1974.

Bordwell, D., Staiger, J. and Thompson, K. *The Classic Hollywood Cinema: Film Style and Modes of Production to 1960.* New York: Columbia University Press. 1985.

Brereton, P. *Environmental Ethics and Film*. London: Earthscan Routledge. 2016.

Brereton, P. *Hollywood Utopia: Ecology in Contemporary American Cinema.* Bristol: Intellect Press. 2005.

Brereton, P. *Smart Cinema: DVD Add-ons and New Audience Pleasures*. Basingstoke UK: Palgrave. 2012.

Buell, L. *The Future of Environmental Criticism: Environmental Crisis and Literary Imagination*. Oxford: Blackwell Publishers. 2005.

Dalbotten, D., Roehrig, G. and Hamilton, P. (eds) *Future Earth: Advancing Civic Understanding of the Anthropocene.* Washington DC and Hoboken NJ: American Geophysical Union and Wiley. 2009.

Devine-Wright, P. 'Beyond NIMBY'ism: Towards an Integrated Framework for Understanding Public Perceptions of Wind Energy' *Wind Energy*, 8: 125–139. 2004.

Estevez-Saa, M. and Lorenzo-Modia, M. J., 'The Ethics and Aesthetics of Eco-caring: Contemporary Debates on Ecofeminism(s)' *Women's Studies: An Interdisciplinary Journal*, 47(2): 123–146. 2018.

Festinger, L. *A Theory of Cognitive Dissonance*. Palo Alto CA: Stanford University Press. 1957.

Friedman, T. *Hot Flat and Crowded: Why the World Needs a Green Revolution and How We Can Renew Our Global Future*. London: Penguin. 2008.

Hawkes, J. 'The Fourth Pillar of Sustainability: Culture's Essential Role in Public Planning'. 2001. Available online at: www.culturaldevelopment.net.au/community/Downloads/HawkesJon%282001%29TheFourthPillarOfSustainability.pdf.

Hiltner, K. (ed.) *Ecocriticism: The Essential Reader*. London: Routledge. 2015.

Hochschild, A. 'Emotion Work, Feeling Rules, and Social Structure' *American Journal of Sociology*, 85: 551–575. 1979.

Hoijer, B. 'Emotional Anchoring and Objectification in the Media Reporting of Climate Change' *Public Understanding of Science (PUS)*, 19: 717–731. 2010.

Hulme, M. *Why we Disagree about Climate Change: Understanding Controversy, Inaction and Opportunity*. Cambridge UK: Cambridge University Press. 2009.

Jenkins, H. *Convergence Culture: Where Old and New Media Collide*. New York: New York University Press. 2006.

Kerr, N. L. and Kaufman-Gilliland, C.M. '… "And Besides, I Probably Couldn't Have Made a Difference Anyway": Justification of Social Dilemma Defection via Perceived Self-Inefficacy' *Journal of Experimental Social Psychology*, 33(3): 211–230. 1997.

Kiehl, T. J. *Facing Climate Change: An Integrated Path to the Future*. New York: Columbia University Press. 2016.

Klas, A., Zinkiewicz, L., Zhou, J. and Clarke, E. J. R. 'Not all Environmentalists are Like That …': Unpacking the Negative and Positive Beliefs and Perceptions of Environmentalists' *Environmental Communication* [advanced copy] 2018.

Lazarevic, D. and Valve, H. 'Narrating Expectations for the Circular Economy: Towards a Common and Contested European Transition' *Energy Research and Social Science*, 31: 60–69. 2017.

Lazer, D., Baum, M. A., Benkler, Y., Berinsky, A. J., Greenhill, K. M., Menczer, F. and Metzg, M. J. 'The Science of Fake News' *Science*, 359(6380): 1094–1096. 2018.

Leiserwitz, A. 'Day after Tomorrow: Study of Climate Change Risk' *Environment: Science and Policy for Sustainable Development*, 46(9): 22–39. 2004.

Lertzman, R. 'The Myth of Apathy' *The Ecologist*. 2008.

Lull, J. 'Family Communication, Patterns and the Social Uses of Television' *Communication Research*, 7(3): 197–209. 1980.

Morgan, M., Shanahan, J. and Signorielli, N. 'Looking Forward, Looking Backward: Ten Questions about Cultivation'. In M. Morgan, J. Shanahan and N. Signorielli (eds) *Living with Television Now*, pp. 389–404. New York: Lang. 2012.

Morton, T. *Being Ecological* Cambridge MA: MIT Press. 2018.

Nabi, R. and Krcmar, M. 'It Takes Two: The Effect of Child Characteristics on U.S. Parents' Motivations for Allowing Electronic Media Use' *Journal of Children and Media*, 10(3): 285–303. 2016.

Roser-Renouf, C., Mailbach, E. W., Leiserowitz, A. and Zhao, X. 'The Genesis of Climate Change Activism: From Key Beliefs to Political Action' *Climate Change*, 125(2): 163–178. 2014.

Schriber, S. and Cole, M. *The Psychology of Literacy*. Cambridge MA: Harvard University Press. 1981.

Slovic, P. and Peters, E. 'Risk Perception and Affect' *Current Directions in Psychological Studies*, 15(6): 322–325. 2006.

Slovic, S. 'The Third Wave of Ecocriticism: North American Reflections on the Current Phase of the Discipline' *Ecozon@*, 1(1): 4–10. 2010.

Stoll-Kleemann, S., O'Riordan, T. and Jaeger, C. 'The Psychology of Denial Concerning Climate Mitigation Measures: Evidence from Swiss Focus Groups' *Global Environmental Change*, 11(2): 107–111. 2001.

van der Linden, S., Malbach, E. and Leiserowitz, A. 'Improving public Engagement with Climate Change: Five "Basic Practice" Insights from Psychological Science' *Perspectives on Psychological Science*, 10(6): 758–763. 2015.

Weik von Mossner, A. *Affective Ecologies: Empathy, Emotion and Environmental Narrative.* Columbus OH: The Ohio State University Press. 2017.

Weintrobe, S. (ed.) *Engaging with Climate Change: Psychoanalytic and Interdisciplinary Perspectives*. London: Routledge. 2013.

3 Promoting new media literacy

Engagement with sustainable environmental literacy

As Gunter Kress affirms in his classic study *Literacy in the New Media Age* (2003)

> it is no longer possible to think about literacy in isolation from a vast array of social, technological and economic factors: new literacies will have profound effects on human, 'cognitive/affective, cultural and bodily engagement with the world and on the forms and shapes of knowledge.

Consequently we need to know more about 'the affordances of each mode of media' (2003: 2).

David Buckingham (2006) posits that media literacy is the outcome of sustained and critical media engagement. Sustainability and environmental education promote green cultural citizenship, which in turn means embodying sustainable behaviours and cultural practices that shape and promote ecological values within the interconnected realms of society, economy and environment. When it comes to sustainability education, one would hope media literacy might encourage various forms of sustainable cultural practices and provide possibly new solutions for the future (Lopez 2013: xvii). It is not enough to be educated about the environment and teasing out how mankind can exist in harmony with it, we also need to understand how people communicate (textually and audio-visually) about the environment and how our feelings about the natural world are ripe for exploitation.

Young people are frequently described as a constituting digital generation, being defined in and through their experience of and with digital computer technology. These attributes are particularly explored in later video games and YouTube chapters. In an early journalistic survey, Lasica (2002) seeks to defend young people from what is seen as 'Hollywood's war against the digital generation'. Elsewhere, 'we encounter "the Nintendo generation" (Green and Bigum 1993), "the PlayStation generation" (Blair 2004), and the "net generation" (Tapscott 1998)', as well as related constructions, such as 'cyberkids' (Holloway and Valentine 2003). Meanwhile, in Japan for instance there has been considerable discussion of the 'thumb generation', involving young people who have apparently developed a new dexterity in their fingers as a result of their (ab)use of game consoles and mobile phones (Brooke 2002, cited in Buckingham and Willett 2006: 1). As recently as June 2018, there has been much publicity around certain video games as officially causing various forms of addiction.

Rather than feeding off moral panic debates around new media, why can't scholars imagine an alternative pro-social and robust 'green generation', or some such niche new digital audience, who in turn are critically engaged with environmental issues? Certainly there is so much research ongoing across the world in trying to evaluate the particular attributes, which might help explain the motivations, mindset and trigger points, driving a multi-faceted new generational audience. Invoking the benefits of pro-social messages, there is much work to be done in explaining how this trajectory might be promoted through more wholesome environmental messaging.

Of course, as Buckingham and many other new media critics affirm, the very idea of a digital generation constantly 'connects those fears and anxieties to technology: it suggests that something has fundamentally and irrevocably changed, and that this change is somehow produced by technology – but should approach these issues with a degree of scepticism' (Buckingham and Willett 2006: 2).[1]

More historically grounded social scientists like Bourdieu (1993) have argued however that (new) generations are more socially and culturally defined. Different generations of course will have different tastes, orientations, beliefs and dispositions (or habitus). For example, if we explore the construction of a popular category such as Generation X, or other new generational cohorts, there is considerable disagreement about their historical parameters, let alone whether the term actually means anything to the people who are allegedly members of this generation (see Buckingham and Willett 2006: 4). Within media and cultural studies, age has (somewhat belatedly) come to join class, ethnicity, and gender as key barometers of social identity. At the same time, more globally prescribed notions of 'green' citizenship and related political agency have a long way to go before this too become normative and a de facto markers of (political, social or cultural) identity, much less a defining benchmark to help regulate the future survival of our planet.

It is inferred, for instance, that while N-Geners are 'hungry for expression, discovery and their own self-development' (Buckingham and Willett 2013, they are savvy, self-reliant, analytical, articulate, creative, inquisitive, accepting of diversity, and socially conscious. Such generational differences are seen to be produced by technology, rather than being a result of other social, historical, or cultural forces. For many from this new generation, Tapscott (1998) argues, 'using the new technology is as natural as breathing' (40). Technology is the means of their empowerment, and it will ultimately lead to a 'generational explosion'. Such self-conscious expressions of 'technological optimism' permeates the scholarly literature and appears to suggest a new form of 'empowerment' for young people, while at the same time reinforcing a generation gap, as the habits and preferences of the older generation appear to be superseded.

But David Buckingham remains critical of this form of blind optimism and uses Tapscott's work primarily as a straw horse to knock such very enticing ideas down, suggesting that the research neglects the fundamental continuities and interdependencies between new media and the old media (such as TV), which Tapscott apparently despises. Certainly, Buckingham is correct in highlighting how a longer historical view is required and clearly shows that old and new technologies often come to converge or at least co-exist over time. Probably nearer the mark, Livingstone *et al.* (2005) conclude that most young people's everyday uses of the internet are characterised *not* by spectacular forms of innovation and creativity, but by relatively mundane forms of information retrieval.

> What most children are doing on the internet is visiting fan websites, downloading music and movies, e-mailing or chatting with friends, and shopping (or at least window shopping). Technology offers them different ways of communicating with each other, or pursuing specialist hobbies and interests, as compared with offline methods, but the differences can easily be overstated.
>
> (Quoted in Buckingham and Willett 2013)

Such an evaluation rings true from everyday observation of the class-room, while communications scholars (probably with too much time on their hands and often striving to promote the radical potential of what they are studying) may sometimes place too large a stake on what has been transformed within the newly mediated technosphere.

Given his relentless optimism, Tapscott and others inevitably ignore the downside of the internet, which Buckingham and Willettt define as encapsulating 'the undemocratic tendencies of many online communities, the limited nature of much so-called digital learning, and the grinding tedium of much technologically driven work'. At the same time, he also calls attention to at least the 'possibility that technology might be used to exploit young people economically, or indeed that the market might not provide equally for all, does not enter the picture' (Buckingham and Willett 2006: 10–11). The technologically empowered cyberkids of the popular imagination may indeed exist. But even if they do, critics like Buckingham are correct in affirming they are in a minority and probably are atypical of young people in general. It is safe to summarise that, 'for most young people, technology is a relatively marginal concern. Very few are interested in technology in its own right, and most are simply concerned about what they can use it for' (ibid.). Hence the role and function of education and environmental literacy in particular, using the affordances of new media technology, should not presume in advance the extra liberating and learning potential of new media formats. Certainly, all of this remains an ongoing debate and poses a challenge into the future, which currently remains dominated by the productive potential of big data and so-called web 2–0 technology.

Generally, the terms 'social media' and 'web 2.0' have become popular for describing types of World Wide Web (WWW) application, such as blogs, microblogs like Twitter, social networking sites, or video/image file sharing platforms like YouTube and wikis. The term web 2.0 was originally coined back in 2005 by Tim O'Reilly, who considered

> the network as platform, spanning all connected devices; Web 2.0 applications are those that make the most of the intrinsic advantages of that platform: delivering software as a continually-updated service that gets better the more people use it, consuming and remixing date from multiple sources.
>
> (Fuchs 2017: 34)[2]

While the study of social media

> describes a specific set of internet-based, networked communication platforms and adapt a business model of a database built by its own users. This potentially enables the convergence of public and private communication includes companies such as Facebook, Twitter, Google, Reddit and Tumblr, Pinterest and Instagram, coupled with the massive growth of Blogger and YouTube. Essentially, social media and related software are tools that at least potentially serve to 'increase our ability to share, to co-operate with one another and to take collective action, all outside the framework of traditional institutions and organisations.
>
> (Shirky 2008: 20–21)

This certainly reflects the utopian potential of such new media technology suggested by Tapscott above. But for the purposes of this study, one must question, as Buckingham and Willett suggest, if these attributes and affordances are actually being fully realised. This book will closely examine a number of new media formats and take into account their capability of promoting new forms of environmental media literacy.

For instance, social media usage is currently being colonised by big data analytics, coupled with new forms of surveillance to become a major business and academic preoccupation that has intensified in times of crisis. Such times of crisis often promote new types of scapegoating in order to distract attention away from deep causes of social and particularly environmental problems (Fuchs 2017: 55). At the same time such data analytics remain an important technological tool to help empirically map the problem and environmental implications of global terrorism for example, as will be illustrated by an analysis of *Homeland* in Chapter 7.

Yet while focusing on the extent of climate change problems across the world, big data co-incidentally through the growth of cloud computing unfortunately leads to further increases in CO_2 waste, while ostensibly promoting quick and easy solution to environmental problems (Fuchs 2017). The consumption of large amounts of (carbon-based) energy needed to maintain data centres and even support cloud storage, coupled with the expediential growth of hardware and consumable digital media technology with its planned obsolescence and short life-span, is greatly adding to new forms of e-waste, which is often dumped back into developing countries, where some of the precious metals used to make the smart phones originated from.[3] Several environmental media scholars (Cubitt 2017; Miller 2011) are rightly calling attention to these waste material scandals. But at the same time, most e-citizens cannot ever imagine, much less contemplate the reality of not having such ever-developing digital information devices that are deemed essential to maintain *our* modern communication age. Consequently, how to measure and evaluate the relative power and effectiveness of such new media affordances, as against their wasteful inefficiencies, remains a major challenge, especially when exploring new forms of environmental literacy. But for the purposes of assessing this tension, this study will nonetheless focus on the growing power of audio-visual media towards hopefully supporting a positive road towards sustainable transformation.

Promoting new modes of active engagement environmental literacy

Sheppard (2012) identifies three simple recommendations for how to achieve a radically new perception of climate change and a greater sense of environmental literacy: (1) make it local; (2) make it visual; (3) make it connected. Beyond actively employing textual analysis and excavating a green trace throughout the total output of popular media, environmental scholarship has to become more preoccupied with the drive to delve deeper into the nuances of audience perception, reception and the creation of more effective emotional stimuli that can actively influence viewers, while at the same time appreciating and understanding audiences' relationships with new forms of on-line media and aesthetics that might in turn support environmental literacy. Extensive analysis around new modes of understanding the effect (and affect) of evolving forms of media are needed to test such assertions and uncover further implications for constructive environmental literacy. But this project involves broad-based and longitudinal audience research across a range of constituencies, which unfortunately remains outside the finite scope of this volume.

Yet a useful start has been made towards understanding audience predispositions towards engaging with environmental issues – at least in film studies – on recalling a number of eco-literary and media publications over the last few years, including most recently Alexa Weik von Mossner's cognitive studies volume (2017), citing Murray Smith's (1995) well regarded 'Engaging Characters' approach to film analysis, which in turn is rooted to a combination of cognitive theory, derived from work by David Bordwell and an emergent application of analytic philosophy. Such multi-purpose approaches can be re-applied to all forms of audio-visual narratives, from documentaries to short Internet tales, alongside long-play televisual series examined in subsequent chapters. Smith's study for instance incorporates a subtle engagement with three basic questions, which also closely relate to unpacking environmental issues and the general representations of nature in fictional narratives:

1. What exactly are 'characters' on screen feeling and how do they relate to persons whom we meet beyond the cinema?
2. What is the role of emotional engagement within media and film?
3. In what ways do the structural organisations of a film call upon us?

(Cited in Barker *et al.* 2001: 3)[4]

Fictional narrative film engagement most certainly exercises all these elements and they can be easily extended onto television and other online media. In particular this analysis includes what characters *feel* about environmental tensions, alongside calling attention to how emotionally charged such stimulus has a hold on audiences. This well-established textual film study approach is applied in this volume to uncovering the inner-life and psychological motivation of televisual characters as diverse as Walter White in *Breaking Bad* or Carrie Mathison in *Homeland*, while at the same time inferring how audiences might connect with and relate to such powerful narratives. Scholars can attempt to tease out how general media and filmic narrative structures, through their overall use of aesthetical strategies, might assist in this long-term process of environmental engagement. Furthermore, applying such strategies to long-play television series, as well as shorter and more interactive multimedia and games stimuli, while assessing their overall impact, remains an ongoing challenge.

Underpinning this investigation also is a desire to examine how media might be read as more ecologically engaging and provocative, while supporting the long-term pedagogical approach of teaching green literacy, using the foundations of an audience research approach. At a purist level, Paula Willoquet-Maricondi (2010) explains that 'eco-cinema', as embodied within polemic documentary stories like *An Inconvenient Truth* (2006), can be defined through films that explicitly express issues of ecological import to further environmental awareness and preservation, while encouraging audiences to reflect on what it means to be part of an eco-system and belong to a global biotic community. Chapter 4, which focuses on food documentaries, illustrates this approach most directly. Unfortunately, if one were to exclusively concentrate on overtly environmentally-driven texts there would be a very limited diet of material to examine, much less enough to drive a pedagogical process of mass engagement around environmental issues. Consequently, the net has to be constantly widened to include a broader range of audio-visual media, which often only tangentially reference environmental issues, either aesthetically and/or through their thematic and narrative content. In any case, such a broader corpus of mainstream media narratives can at least help to keep environmental issues firmly in the public domain, while also being useful in suggesting new modes of critical engagement.

While a growing body of environmental analysis is concentrated on the semiosis of the text, usually focusing on conventional documentaries or fictional films, little sustained scholarship has yet been carried out on a comparative analysis of audiences' perceptions and interpretations in demonstrating how mass media's cognitive and emotional connections might be effectively registered or codified by viewers. In recent decades, audience reception studies have emphasised the active, interpretative, critical, creative and sometimes even the resistant nature of public engagement with the media. Yet, while most film and media analysis has made claims about the audience, these claims are seldom made explicit or empirically examined. Linking such audience research modalities within a general discussion of environmental textual literacy, certainly helps to create a fruitful and engaging cross-comparison and juxtaposition.

Critics for instance observe that meaning-making is 'best understood as a set of practices in and through which, either consciously or unconsciously, human beings "make sense" of themselves and their world' (Haines-Lyon and Marsh 2007: 115, cited in Lyden 2009). This sense-making is integrated into how we live:

> [P]eople's structuring of their lives, and the shaping of their life-practices (e.g. how they earn and spend money, how they use natural resources, how they conduct their personal relationships, how they spend their leisure time, whether and how they value family ties etc.) are all informed and worked out, in conversation with the texts and practices of popular culture.

In other words, popular mediated texts including online advertisements, as explored in a case study at the end of this chapter, can be decoded as a portal or window into audience perceptions and even help to explain preferred trigger points around behaviour change patterns.[5]

As further affirmed across later chapters, in order to facilitate and enable such meaning-making to occur all forms of media in various ways invite affective responses on the part of the viewer to promote attentiveness, which in turn 'gives pleasure, stimulates the imagination, and provides a communal context in which the viewing can occur' (Marsh 2007: 69–71). Essentially, 'visualising the environment in order to comprehend it is a constitutive aspect of making the environment meaningful' (Doyle 2009: 285). Ever since the Enlightenment, this combination of visual aesthetics and epistemology has been activated and inscribed within landscape painting, photography, and more recently digital and even satellite images (Ingold 1992). Audio-visual media can further push such two-dimensional visualising and news dissemination, towards a more immersive form of three-dimensional engagement, promoting new forms of environmental experience. Such visualisations remain a primary focus for examining the mediation of ecological debates in particular and for exploring their effects on audiences and citizens generally. It is often asserted, especially within the development of visual literacy, that notions of place remain the most immediate and visceral manifestation of environmental engagement.

Recalling Sheppard's (2012) three simple recommendations for how to achieve a radically new perception of climate change and to promote a greater sense of environmental literacy, climate change communication further becomes more salient by bringing relevant information down to the local level, putting it into a social context that people care about. For example, consciously using the local landscape to raise and express climate change issues and thereby actively engaging citizens towards developing local solutions (cited in Schroth *et al.* 2014: 5). The key question for environmental media, and all potentially transformative arts for that matter, is finding a way to highlight and codify such salient features, so that they speak more effectively to a wider cohort of audiences than simply those already predisposed to accepting and embracing the cautionary environmental message. Consequently, the broader communication process and overall strategy has to be more all-embracing and the allegorical use of metaphors and what I call the creative imaginary of the media production process, has to be both provocative and enticing, even if only tangentially focused around specific green issues. All the while, at an educational level, highlighting a primary pedagogical principle in suggesting that audiences have to learn to critically decode texts, using a range of environmental strategies, while learning to feel confident in decoding the audio-visual stimulus always remains an ongoing challenge.

Again, while long term audience analysis is needed to trace and test the adequacy and effectiveness of green literacy programmes, Social Representations Theory (STR), for example, appears to be a useful starting point in helping to turn the unfamiliar into the familiar. This is a theoretical paradigm concerned with how the public make sense of the unfamiliar and how they transform scientific concepts like those around climate change into common sense expressions. STR has been used to study public understanding of climate change (Smith and Joffe 2013), and is neatly constituted through two main processes: anchoring i.e. 'the attempt to settle a new, and therefore strange, meaning into the established geography of symbols of a community', and objectification, i.e. giving a new object of knowledge 'a concrete, almost "natural" face' (Jovchelovitch 2001: 173, cited in Marcu *et al.* 2015: 549). While not having the resources to test such a hypothesis in this volume, the science of climate change communication certainly needs to develop its ongoing work by articulating the cultural appeal and pleasures of all types of generic fare, modulated across a growing number of platforms and media outlets, which can in turn assist mass audiences in this process of global environmental engagement and transformation.

As explored in later chapters, time spent by new audiences viewing online videos on YouTube, Vimeo, Hulu, as well as on Netflix, is growing exponentially. New generational media consumers are likely to be choosing their viewing content more noticeably than a decade ago, by turning to the internet as a massive archive and database to source and watch their favourite shows, independently of when they were aired in the first place, much less accessing their news and educational media within pre-set and time-based modalities. Time shifting of media and atomised niche audience consumption is becoming the norm. It would appear that teenagers and students in particular spend countless hours on Netflix and YouTube watching short movies and surfing the internet, which hopefully include climate change stories. In a recent study by Dudo *et al.* (2011), such internet use was deemed a significant predictor of positive science attitudes, directly and through an increase of scientific knowledge (Morgan *et al.* 2012). In any case, more science and environmental media output needs to be developed with targeted stimuli and allegorical markers to entice new generational audiences to engage with an evolving media landscape and an ever-expanding growth across digital platforms.

Promoting more allegorical/metaphorical images of nature and the environment

To combine often conflicting trends across the broad range of environmental literature in itself poses a challenge. The historicising of the wilderness concept, most notably promoted by the environmental historian William Cronon, is undoubtedly one of the most important critiques within the evolving cannon of environmental criticism.[6] At the other extreme, somewhat provocatively as is his forte, Timothy Morton asserts that 'remaining in cynicism is a habit of the beautiful soul. Our choice is false as it has been reduced to one between hypocrisy and cynicism, between wholeheartedly getting into environmental rhetoric and cynically distancing ourselves from it.' Morton goes on to suggest that in 'serving up lashings of guilt and redemption, might ecological criticism not engage the ideological forms of the environment, from capitalist imagery to the very ecocriticism that opposes capitalism?' (Morton in Hiltner 2015: 237).

Meanwhile, Dana Phillips' *The Truth of Ecology* (2003) lambasts environmental scholars for adhering to an obsolete notion of ecological science and for transferring ecological terms to literary study by means of mere metaphor (42–82). Nonetheless, this study seeks to explore how new forms of online metaphors and allegories, which are privileged across a wide range of audio-visual narratives, can be regarded as more actively speaking to environmental debates, while at the same time being cognisant of such concerns. Phillips is probably right in cautioning ecocritics against undue metaphorising, moralisation or spiritualisation in their use of scientific concepts and in calling for more up-to-date knowledge and engagement with scientific literacy. Such new forms of high-level scientific literacy would minimally require some training in quantitative methods and data analytics skills that unfortunately is often lacking across mainstream Humanities education.

Alternatively, however, while addressing mainstream mass communications, it can be counter-argued, *more* metaphors are needed not less in helping to counteract the sometimes complex and rarefied scientific literature of Environmental studies 2.0. Mass audiences, especially those learning from an environmental perspective, need the stimulus of broad-based active engagement in the first place, so as not to be so easily turned off by the density and complexity of the science. Hence espousing the critical heights and nuances of academic and scientific rigour is probably less effective in achieving tangible results with mass audiences in the real world. Pedagogically, at least, scholars have to be constantly willing to compromise and tease out such inherent complexity, while reconstructing tangible and engaging (metaphorical) examples to explain and illustrate the broad mesh of interconnecting environmental sciences and clearly uncover its effects on the planet. Therefore, environmental literacy as a process and model for communication has a long way to go to achieve global traction and needs every assistance it can draw on. All the while recognising there is no magic bullet formula to satisfy the gaping chasm of environmental knowledge, but at the same time calling on audience active engagement to assist in this struggle.

Looking back on Rachel Carson's influential indictment of pesticide overuse in *Silent Spring* (1962), we can also learn a lot from her skilful use of metaphors, especially using the pastoral, biblical apocalypse, nuclear fear (in her comparisons of chemical contamination with radioactive fallout), and 1950s anti-communism ('a grim spectre has crept upon us almost unnoticed' (3)). By all accounts, this sophisticated allegorical and mythical strategy is frequently re-used to powerful effect in eco-film analysis (see Brereton 2005/2012/2016), while also being clearly evident in the (over)romanticism of environmental events, as illustrated in the audio-visual exemplar *An Inconvenient Truth* and its less successful 2017 sequel discussed in Chapter 5. Meanwhile, we now turn to environmental debates which have been co-opted by greenwashed modes of media engagement and a long-established romantic myth around the priorities of landscape and farming. This exploration hopes to sow the seeds of more productive forms of reflexive pedagogical engagement.

Advertising and greenwashing: promoting new modes of production/consumption

Many environmental scholars worry not just about getting environmental messages across, but also about the pervasive nature of (eco)consumerism, which consists of 'greenwashing' (commodifying nature to produce capital) and 'greening' (commodifying nature to produce capital to be used in the conservation of nature). All of these related concerns pose an impossible question: can some form of (pure) environmentalism and green-learning participate in the commodity system, without being intentionally or otherwise co-opted by it (see Lebduska 2003)?

Historically for radical ecologists, the first border laid down is the ideological boundary between green (lowercase) and Green (uppercase). As Andrew Dobson explains, back when the discipline was being established, 'green' was often characterised as shallow and constituting a form of light green ecology, which in turn signifies environmentalism and its attendant acts, such as recycling, attaching catalytic converters to cars and appending scrubbers to industrial smokestacks. By contrast, Green, also known as deep or dark green ecology, signifies ecologism, 'a political ideology in its own right' (Dobson 1990: 5), which alternatively calls for substantial political structural changes in the way resources are used and at its most radical, promoting a form of de-growth, alongside legitimating the growing popularity of the so-called 'circular economy'.

While many (technologically determinist scholars) believe that the major transformation needed to address the growing dilemma of climate change will not be adequately faced up to in time, hence the fall-back, non-revolutionary strategy will constitute some form of complicit 'business as usual' type of adaptation. Thereby applying some manageable levels of mitigation and most likely appealing to a radical turn towards geo-engineering or other scientific solutions to stave off climatic disasters. Such tensions and possible scenarios for effective solutions can be characterised as, 'the environmentalist consumes differently; the ecologist consumes less'. Because ecologism 'contests consumption itself, it remains ideologically opposed to capitalism, unlike environmentalism which often exists comfortably, even collusively, within a capitalist frame' (Lebduska 2003: 144).

Nonetheless, such crude yet clear differentiations in defining and naming does not carry through into the evolving literature, with a wide variety of environmental and ecological definitions being often used indiscriminately and for varying purposes. This volume certainly does not use environmentalism and ecology in such a crude and binary manner. Meanwhile, even strongly invested environmental NGO's like Greenpeace are not so pure, as is evident within their corporate praxis. By cataloguing for instance 'selling green images of animals', this strategy remains complicit with its 'light green' agenda (Lebduska 2003: 144). But even the possibility of a (realistic or pure) media alternative to such complicit and steady-as-she-goes environmental forms and mode of evolution, is always set in opposition to ostensibly more radical forms of revolution. Mediated environmental issues constantly play across both polarities and consequently examination and definitional analysis can never be deemed such an exact science.

At a pragmatic level of human engagement, ecoconsumerism, for instance, assumes a variety of guises, using adverts and other mediated texts to create natural images for products, to signify environmental harmlessness and to proclaim the ever-increasing necessity for environmental activism. Unfortunately, quite often greening and greenwashing have become indistinguishable, as both create particular codes and involve consuming objects and services (Lebduska 2003: 147). Such elisions and ideological valorisations of 'business as usual' modalities need to be always highlighted and called attention to, across any critical environmental literacy project.

For example, in a famous case study by Robin Anderson titled 'The Crying Indian: Corporations and Environmentalism: A half-century of struggle over Environmental Messages', this iconic historical image is regarded as being the most effective advert ever made in promoting a renewed form of environmentalism. Frequently however such messages are designed to obscure, hide and redefine realities, which in turn remains the uncontested definition of greenwashing (Anderson in McAllister and West 2015: 407), unlike the ugly pollution that caused Iron Eyes Cody to cry, 'the version of nature brought to you by today's corporate imaging is blissfully pristine and uncontaminated, unspoilt by over-extraction and waste.' Such tensions are illustrated and evidenced in the farming advert to be discussed presently.

Meanwhile, Toby Miller's (2017) *Greenwashed Culture* is most vociferous in critiquing Hollywood and its greenwashed attempts to speak to environmental issues 'through spectacle and science alike, to convince the world's public of the reality of our ecological situation' (1). Furthermore, as Sean Cubitt, Miller and others point out, media literacy educators have an ongoing responsibility to address environmental issues wherever they can, especially as it is becoming better known that mass media and educational technology as an industry in itself, contributes extensively in their own right to Co2 emissions and general environmental pollution (see Cubitt *et al.* 2011: xix; Cubitt 2017).

Nevertheless, being both pragmatic and proactive and in seeking to discover some positives to help encourage the media, alongside the educational industry to strive to uncover new modes of address – while also embracing more critical media/environmental literacy strategies – environmental scholars probably should try to unpack popular illustrations and examples (even if somewhat tainted, even greenwashed) to help shine more positive light on the enormous challenges ahead.

This multi-pronged strategy is evidenced through the following case study reading of an online advert for Irish farming, portraying some fantastically beautiful if greenwashed images of land and agriculture. Notably, in the last 30 years, even in the midst of green consciousness and recycling rhetoric, the mountains of garbage across all areas has doubled. Such unsustainable developments can be traced to waste streams and to the Post-War industrial practices of planned obsolescence and the ever-present legitimation of an over-consumptive, throw-away culture. The same practices incidentally that give birth to the need for the Crying Indian advert. Advertising visions of virtual nature in harmony with consumption frequently hide the long-term destruction of the environment (Anderson in McAllister and West 2015: 416).

To help illustrate some of these polarising tensions, later chapters will address new innovative forms of media and representations of this exploration of waste and other environmental issues, while this chapter will conclude with an examination of an Irish advertising campaign for farming and food production, which can at the outset be accused of a crude form of greenwashing. The following case study further serves to highlight a range of semiotic tools that can be used for critical analysis, which alternatively can assist in highlighting environmental literacy protocols and strategies for audiences to engage with. This case study certainly serves as a useful pedagogical exemplar for engaging with new forms of active environmental literacy.

Irish case study: green marketing – www.origingreen.ie

Bord Bia's (a public body in charge of promoting milk production and consumption in Ireland) ongoing audio-visual advertising campaign uses drone aerial photographs of fantastic but stereotypical touristic Irish green landscapes that appear so symmetrical and beautiful, as the land is tracked from above through a filter of clean fresh air. This aesthetic strategy of framing landscape echoes earlier advertising campaigns featuring the Irish film star Saoirse Ronan (www.ireland.ie/en/stories/food-thought-ireland-goes-green) – which is reminiscent of the Russell Crowe blockbuster *Gladiator* (2000) – as Ronan stretches out her hands and 'feels the grain' on her finger tips, while walking through the lush cereal crop fields, spouting evocative patriotic and green (washed) rhetoric. As the Irish musical motif builds up, viewers admire this rising Hollywood film star (a great stroke of casting to use a relatively young woman to speak to and for farming), walking through the lush fields, as she emotionally recites the powerful call to support Irish agriculture and 'put on the green jersey', thereby supporting a deep form of national identification and by extension extolling a comforting form of environmental sustainability. 'We did not inherit this world from our parents, we borrowed it from our children' continues in this (environmental first wave or 'Old Testament') vein that 'sustainability is what the world needs now' and confirms most clearly how it 'is the defining opportunity of our time and Ireland can become a world leader'.[7]

In response to such evocative if greenwashed advertising, which was aired on national television and across the internet, environmentalist and activist John Gibbons, who contributes regularly to the *Irish Times* (being the 'newspaper of record' and who has his own useful website www.thinkorswim.ie),[8] was particularly scathing in his attack at such blatant modes of greenwashing. 'Bord Bia's (Irish Food Board) latest lush, evocative "Origin Green" advertising campaign, which ironically opens by panning from a height across seemingly pristine rivers and bucolic pastoral scenes is reminiscent of a Constable painting'. Yet all this celebration of landscape, using ubiquitous drone imagery, can and should be read within a context of the recent Environmental Protection Agency report on Water Quality in Ireland (2010–2015), which illustrated serious pollution levels. This situation was aggravated no doubt because of farming effluent of all types, as the country strives to increase productivity across the island. Farmers, as highly-regarded stewards of the land, as against being despoilers of their inheritance through over-production and excessive use of fertilisers, have become the focus of critical ire for many (urban) environmental activists and organisations, together with *An Taisce* (Irish Environmental Protection Organization). Such tensions are especially prescient as the country has to deal with increased levels of methane (CO_2) production, through the expediential growth of Ireland's beef and dairy herd.

Incidentally, to further aggravate critical media disquiet, the advertising campaign was awarded to the Rothco advertising company and came in with a huge budget of over €536,000. Furthermore, Gibbons rightly wonders who the advert is aimed at and what its overall purpose might be? Critical questions which environmental media analysis constantly poses. On contacting Bord Bia to estimate what restrictions were used to assess the farmer's quality assurance in achieving this sustainability kite mark, it was confirmed that almost all farmers were automatically included as being de facto 'sustainable'. Recalling similar greenwashed adverts by for instance Bord na Móna (state body who are in charge of developing and exploiting Irish nationalised boglands), one could easily agree that this 'blizzard of eco-blarney' is serving no useful purpose, except as a means to promote Ireland as a (touristic) green agricultural and energy island abroad and is somewhat reminiscent of the current co-opting of the mythic, monastic site and bird sanctuary island of Skellig Michael for the *Star Wars* franchise.

In the same vein, *The Irish Times* report by their ex-editor and environmental correspondent Kevin O'Sullivan (4 October 2017), repeats this critical engagement, while using a provocative headline 'Origin Green is a Sham' and strongly suggests, like Gibbons earlier, that this official certificate is simply a marketing tactic and a further example of vacuous greenwashing.[9] Using emotive PR language, the website (www.origingreen.ie) consciously sets out what 'being green' means and how 'Origin Green' is about measuring and improving how *we* do things: especially for farmers. Bord Bia has been auditing and certifying good farming standards in food safety, traceability and animal welfare for over 20 years through its quality assurance scheme. They also promote responsible farm management measures and support biodiversity and the efficient use of water, energy, feed and fertiliser, all of which remain its primary objective. All of course good and valuable environmental interventions for developing best practice models across all aspects of agriculture.[10]

However environmental critics like Gibbons or full-time environmental journalists like O'Sullivan state that such an approach simply legitimises a 'business as usual' strategy.[11] Motivation for all of this marketing activity appears to be focused around producing good/safe food in the most sustainable way possible. By all accounts nutritious food simultaneously protects and enhances the natural environment and the local community. Essentially, this means embracing the needs of the present, without compromising the future. Consequently, in all the general guidelines of the organisation, there appears to be little or no suggestion that methane emissions from a growing dairy herd needs to be controlled, or for that matter that more regulations are required to ensure long-term sustainability benchmarks.

The website goes on to affirm how 'Ireland is unique', with the climate supporting a relatively long grass growing season, having plenty of natural precipitation to maintain a robust agriculture eco-system. Essentially, it is affirmed that a grass-based cattle system *is* more efficient, and more environmentally sustainable than intensive indoor annual feeding systems. But this is not what is being consciously promoted by official Government and Agricultural policy, as set out within the current policy document 'Harvest 20/20'.[12] This official policy actively supports increasing dairy herds and the growing production of animals for slaughter and thereby uncharacteristically seeks to promote more factory-type sheds and grain supplements. All this is needed to speed up the growth process and provide an increased return on investment, while ensuring that the animals spend less time – especially during the long winter months – on the fragile soil and grassland. Such a clearly articulated factory-type agricultural growth strategy unfortunately does not concur with a nurturing and holistic farming model, which is further illustrated by such romantic outdoor imagery.

Nonetheless, such adverts remain evocative and potentially educational I suggest, while acknowledging the pervasive danger of a green-washing effect, as enunciated by Gibbons and O'Sullivan above. Environmentally literate critics ought to be able to critique, as well as appreciate and hopefully turn around, such potent adverts on their head and even suggest a new form of radical engagement. At the same time, educationalists have to be always open to examining if audiences actually learn to read in more positive ways through appreciating a number of productive environmental messages, around good land husbandry.[13]

Transforming greenwashed adverts into creative environmental discourses

Such alluring and potent imagery of land and agriculture at least calls attention to the importance of taking into account environmental modes of farming and food security, even if not necessarily agreeing with their economically-driven solutions. Nonetheless, if re-filtered through a process of environmental media literacy, as laid out across this and other chapters and through critical re-framing of such discursive practices, a more positive eco-reading can be made. Such re-purposing can serve to actively encourage audiences to more positively engage with the environmental challenges we face. At the very least, such adverts and filmic representations can keep an environmental agenda on the table for mass audiences, which in itself is a useful exercise. All the while recognising comments by Miller (2017) and other scholars with their well-argued political and ideological criticism of the inherent dangers of greenwashing. Or acknowledging the longer critical history of romantic landscape art and literature practice, which is usually explored within environmental/ideological media studies courses to help decode their ideological normalising of pro-consumption messaging. Alternatively, such iconic and powerful stimuli can be re-purposed to help counteract their dominant or preferred messaging. As laid out by numerous cultural media scholars, most notably the late Stuart Hall, negotiated or even oppositional readings can be used to help uncover pro-environmental readings and interpretations.

Active critical strategies can greatly be assisted at a practical level through for instance mashing-up such adverts (see www.thinkorswim.ie) and analysing how they might alternatively encourage students to produce their own creative/radical responses to food production and consumption. Media students, in particular, can use their ever-expanding practical, technical and creative skills to creative effect. Users and audiences generally can be assisted in teasing out alternative meanings and counter-readings – as for instance feminist media scholars have carried out over the years with regards to conventional patriarchal media outputs. In other words, instead of always bemoaning the lack of constructive and critical environmental media to engage with, or simply dismissing so much populist artefacts out of hand, the environmental communications fraternity might more actively focus on the media that already exists across so many platforms. By helping to demonstrate how such output might be re-evaluated and re-constituted from a range of environmental contexts, rather than simply discussing their well-worn one-directional ideological agenda, remains a more fruitful pedagogical agenda. Of course, such efforts constitute a risky and difficult educational strategy, but one which needs to be actively worked on over time. This process can for instance be applied using a wide range of examples that speak to different cohorts of students (alongside the general public) and across the bread and butter research aims and strategies laid out by media literacy programmes. Such an open-ended approach can further help construct more fruitful avenues for developing robust environmental literacy strategies and encourage a more effective pedagogical environmental agenda, while constantly highlighting potentially new avenues for producing counter-narratives that can be applied across the growing new media landscape.

Powerful creative imaginaries, including such opaque greenwashed adverts, as well as all other forms of media explored in subsequent chapters, can and should be applied as both a stimulus and raw material for pedagogical purposes. The adverts and mass media generally can be re-imagined or reconstituted to promote more productive, even transgressive forms of environmental engagement. Greenwashing, nostalgia, pastiche and irony, alongside promoting active engagement with romantic representations of landscape across various artistic formats, are most certainly useful across the evolving media landscape and are necessary to engage with across all stages in the process of learning to become more critically engaged with old and new manifestations of environmental issues. Using easy examples like these adverts, recalling environmentally sensitive farming practices can assist in an ever-expanding learning process.

Simply categorising or dismissing images as romantic, regressive or 'anti-environmental', while setting up clear binaries between environmental and ecology or first and second wave environmental readings is not finally helpful for a critical educational process in the long term. There are few enough types of images available to engage with, much less memorable narratives that highlight cogent analysis of environmental concern that communicators probably should not dismiss out of hand those that don't appear to fulfil a pre-given expectation and therefore are unconsciously dismissed as a form of greenwashing. Consequently, media scholars and environmental activists should alternatively adapt and reapply a broad range of potent signifiers, including popular audio-visual adverts as well as the full gamut of commercial and popular media output, which audiences enthusiastically connect with, while striving to widen their reflective and critically pedagogical practice.[14]

Finally, of course echoing a '101-media literary studies' pedagogical approach, environmental and educational scholars should always be cognisant of how semiotic signs and images can provoke multiple (polysemic) meanings and readings. This can therefore be applied to critique the preferred message of the advertiser, including as suggested earlier, subscribing to an unsustainable environmental farming model for the future. Much creative effort and resources are of course used to produce advertisements, alongside fictional televisual, film and other online narratives. Their ultimate success speaks to audiences' deep engagement with such expressions of local and national place engagement. Hence pro-environmental groups who often strive to promote an alternative agenda, ought to find ways of re-viewing and re-purposing such compulsive and engaging advertising strategies and texts for their own ends and in turn help to develop effective new modes of engaging with and promoting cogent environmental literacy protocols.[15]

Notes

1 This debate is reminiscent of Donna Haraway and other science fiction cultural theorists trying to re-imagine feminist discourses, while drawing on cyborg studies of post-human agency and prosthetics.

2 Michael Mandiberg helpfully argues that the notion of 'social media' has been associated with multiple concepts: the corporate media favourite 'user-generated content', Henry Jenkins's media-industries-focused 'convergence culture', Jay Rosen's 'the people formerly known as the audience', the politically infused 'participatory culture', Yochai Benkler's process-oriented 'peer-production', and Tim O'Reilly's computer-programming-oriented 'Web 2.0' (Mandiberg 2012: 2, cited in Fuchs 2017: 37).

3 In 2012, for instance it was estimated data centres used electricity that equalled the output of 30 nuclear power plants and this dramatically increases with greater usage of online media services. Also, data centres tend to use diesel generators as back-up power supply system's (Fuchs 2017) and produce pollutants that are released into the air and the soil.

4 Drawing on various avenues of psychological research around audience studies, possible models for appropriate questionnaires might include the following, drawn from work of Martin Barker, to help tease out variations around notions of 'identification' or 'empathy', with regards to a study of eco-affect in film.

   1 I tried to understand the characters in the movie by imagining how things looked from their perspective
   2 I really got involved with the feelings of the characters in the movie
   3 While watching the movie, I easily put myself in the place of one of the leading characters
   4 While watching the movie, I felt as if the characters' thoughts and feeling were my own
   5 While watching the movie, I imagined how I would feel if the events in the story were happening to me
   6 While watching the movie, I tried to imagine what the characters were thinking
   7 I became very involved in what the characters were experiencing throughout the story
   8 While watching the movie, I experienced many of the same feelings that the characters portrayed.

   [scale 1=Strongly disagree and 5= Strongly agree. Number in parentheses indicates the Cronbach's alpha coefficient. The empathy was measured only to the experimental group with the control group not measured.]

   (Hyounggon, Kim *et al.*, cited in Barker *et al.* 2001)

5 While not as 'direct' as the ongoing political scandal of appropriating millions of Facebook profiles of real online users, using data analytics and a range of personality questionnaires to assess users' ideological inclinations and attributes, which have facilitated micro-targeted political messaging across a number of political time-frames and events. Again the 'bank of screens' to produce 'Fake News' as visualised in *Homeland* comes to mind as an effective image.

6 While at another end of the spectrum, a small number of ecocritics, such as Joseph Carroll and Glen Glov would like to make the life sciences in general and evolutionary theory in particular the foundation of literary study, recalling E. O. Wilson's idea of 'consilience'. This group seeks to explain cultural phenomena in terms of what they

accomplish for human adaptation and survival. Many scholars in the humanities however, almost instinctively recoil in horror from such a deterministic sociobiological agenda, associating it with pernicious forms of social Darwinism or even a form of Nietzschean ideology, which has historically helped to legitimate totalitarian political ideologies.

7 Available online at: www.ireland.ie/en/stories/food-thought-ireland-goes-green.

8 This daily Irish newspaper is recognised as the 'newspaper of record' and Gibbons has his own useful website www.thinkorswim.ie and www.thinkorswim.ie/i-have-fed-and-starved-species-greater-than-you/.

9 Some of the country's worst polluters include Arrabawn in Tipperary and Dairygold Co-op in Mallow Co. Cork. Yet Origin Green claims to be the only sustainable programme in the world that operates on a national scale, uniting government, the private sector and food producers through Bord Bia. Bord Bia incidentally received over 32.5 million euro in public money for marketing of Irish food in 2016, in contrast with The National Park and Wildlife organisation, the body which manages National Heritage, only receiving 13.5 million euro in 2015.

10 Consequently, they set out to inspect what they designate as 'green farms' every 18 months and compile data on the sustainability of each unit. But the organisation remains simply a marketing advisory body, unfortunately having no teeth, designed to admonish those not following 'good practice'. Instead, they allow working farmers to make their own informed decisions concerning how they might increase the sustainability of their farms.

Note, three Origin Green companies appeared on the EPA's 'worst polluters' list https://greennews.ie/three-origin-green-agri-businesses-named-epas-worst-polluters-list/.

11 There are nonetheless over 50,000 farmers in the newly launched sustainable beef and lamb assurance scheme (over 90% of all farmers) and over 15,000 (85% of all dairy farmers?) in the sustainable dairy assurance scheme.

With regards to business, Bord Bia seeks to mandate – or at least strongly suggest – the maximum of areas of consideration including: raw material and sourcing, manufacturing processing and preparations together with energy, water, waste, emissions and biodiversity. Social sustainability, at its core includes health and nutrition, community initiatives, and employee wellbeing.

12 There has since been an update – Food Wise 2025 – available online at: www.agriculture.gov.ie/foodwise2025/.

13 Calling to mind the current spectre of Brexit and threats to Irish agricultural production – while still appreciating such well-argued critiques of such complicit images. The precarious nature of the agricultural business is further alluded to by the growing dependence Ireland has on exports, with severe worries at present over what will happen with Brexit and Great Britain leaving the EU market, while Ireland remains a full member. Almost 90% of total Irish food and drink are ostensibly produced for export and much of this is sold to our nearest neighbour. Naturally this causes major concerns, considering the volatility of the international markets and noting ongoing currency adjustments.

14 Especially choosing audio visual images and stories, as explored through this volume, which teachers might at first instinctively object to for their blatant ideological evocation of 'business as usual' marketing of nature etc., ought to be taken on board, even risked for the greater potential gain involved.

15 As most provocatively illustrated by the mash-up of Food and Energy advertisements initiated by John Gibbons, very innovative use of a little girl on the swing is used, as a potent repost to the dominant advertising messages and thereby help to kick start some critical engagement.

References

Barker, M., Arthurs, J. and Harindranath, R. *The Crash Controversy. Censorship Campaigns and Film Reception*. London: Wallflower Press. 2001.

Brereton, P. *Environmental Ethics and Film*. London: Routledge. 2016.

Brereton, P. *Hollywood Utopia: Ecology in Contemporary American Cinema*. Bristol UK: Intellect. 2005.

Brereton, P. *Smart Cinema, DVD Add-Ons and New Audience Pleasures*. Basingstoke UK: Palgrave. 2012.

Buckingham, David and Willett, Rebekah (eds) *Digital Generations: Children Young People and New Media*. Mahwah NJ: Erlbaum Publishers. 2006.

Cubitt, S. *Finite Media: Environmental Implications and Digital Technologies*. Durham: Duke University Press. 2017.

Cubitt, S., Hassan, R. and Volkmer, I. 'Does Cloud Computing have a Silver Lining' *Media Culture and Society*, 33: 149–158. 2011.

Dobson, A. *Green Political Thought: An Introduction*. London: Unwin Hyman. 1990.

Doyle, J. 'Seeing the Climate? The Problematic Status of Visual Evidence in Climate Change Campaigning'. In S. Dobrin and S. Morey (eds) *Ecosee: Image Rhetoric Nature*, pp. 279–298. New York: State University SUNY Press. 2009.

Dudo, A., Brossard, D., Shanahan, J., Scheufele, D., Morgan J. and Signorielli, N. 'Science on Television in the 21st Century: Recent Trends in Portrayals and Their Contributions to Public Attitudes Toward Science' *Public Understanding of Science*, 36(6): 754–77. 2011.

Fuchs, C. *Social Media: A Critical Introduction*, 2nd edition. London and Thousand Oaks CA: Sage. 2017.

Hammond, P. *Climate Change and Post-Political Communication: Media, Emotion and Environmental Advocacy*. London: Routledge. 2018.

Hiltner, K. (ed.) *Ecocriticsm: The Essential Reader*. London: Routledge. 2015.

Holloway, S. and Valentine, G. *Cyberkids: Children in the Information Age*. London: Psychology Press. 2003.

Ingold, T. 'Culture and the Perception of the Environment'. In E. Croll and D. Parkin (eds) *Bush Base: Forest Farm. Culture, Environment and Development*, pp. 39–56. London: Routledge. 1992.

Johnson, D. 'Activating Activism: Facebook Trending Topics, Media Franchises, and Industry Disruption' *Critical Studies in Media Communication*, March 2017.

Kress, G. *Literacy in the New Media Age*. London: Routledge. 2003.

Lebduska, L. 'How Green was my Advertising'. In M. Branch and S. Slovic (eds) *The ISLE Reader: Ecocriticism, 1993–2003*, pp. 143–154. Athens GA: University of Georgia Press. 2003.

Lasica, J. D. 'When Bloggers Commit Journalism' (September 2002).

Livingstone, S., Bober, M. and Helsper, E. 'UK Children Go Online: Final Report of Key Project Findings'. London: London School of Economics and Political Science. 2005. Available online at: http://eprints.lse.ac.uk/397/1/UKCGOonlineLiteracy.pdf

Lopez, A. R. 'Greening the Media Literacy Ecosystem: Situating Media Literacy for Green Cultural Citizenship'. PhD thesis submitted to Prescott College in Sustainability Education 2013.

Lyden, J. *The Routledge Companion to Religion and Film*. London: Routledge. 2009.

Marcu, B. A., Gaspar, R., Rutsaert, P., Seibt, B., Fletcher, D., Verbeke, W. and Barnett, J. 'Analogies, Metaphors, and Wondering about the Future: Lay Sense-making around Synthetic Meat'. *Public Understanding of Science*. 2015.

Marsh, C. *Theology Goes to the Movies: An Introduction to Christian Thinking*. London: Routledge. 2007.

McAllister, M. and West, E. (eds) *The Routledge Companion to Advertising and Promotional Culture*. London: Routledge. 2013/15.

Miller, T. *Greenwashed Culture*. London: Routledge. 2017.

Morgan, M., Shanahan, J. and Signorielli, N. (eds) *Living with Television Now: Advances in Cultivation Theory and Research*. New York: Peter Lang. 2012.

Mosco, V. 'Marx in the Cloud'. In Christian Fuchs and Vincent Mosco (eds) *Marx in the Age of Digital Capital*, pp. 516–535. Leiden: Brill. 2016.

Phillips, D. *The Truth of Ecology*. Oxford: Oxford University Press. 2003.

Preston, P. *Reshaping Communication, Technology, Information and Social Change*. London: Sage. 2001.

Schroth, O., Dulic, A. and Sheppard, S. 'Visual Climate Change Communication: From Iconography To Locally Framed 3D Visualization' *Environmental Communication*, 8(4): 413–432. 2014.

Sheppard, S. R. L. *Visualising Climate Change: A Guide to Visual Communication of Climate Change and Developing Local Solutions*. London: Routledge. 2012.

Shirky, C. *Here Comes Everybody*. London: Penguin. 2008.

Smith, N., and Joffe, H. 'How the Public Engages with Global Warming: A Social Representations Approach' *Public Understanding of Science*. 2013.

Tapscott, D. *Growing Up Digital: The Rise of the Net Generation*. New York: McGraw Hill. 1998.

Weik von Mossner, *Empathy, Emotion and Environmental Narratives*. Columbus: Ohio State University Press. 2017.

Wodak, J. 'Shifting Baselines: Conveying Climate Change in Popular Music' *Environmental Communication*, 2(1): 58–70. 2018.

Wurtzler, Steve J. *Electric Sounds: Technological Changes and the Rise of Corporate Mass Media*. Berkeley CA: California University Press. 2007.

4 Food documentaries and green anxieties

Actively promoting environmental literacy

Overview

This chapter will explore how popular and long-established environmental documentaries through their use of direct address and creative aesthetics and imaginaries describe a range of cautionary tales that speak to the environmental importance of food production, waste and (over)consumption. The central problem for ethical food production is that it appears politically unacceptable to get across complex environmental and ethical messages around food security, much less acknowledge that the global meat-based economy ought to shrink, while struggling to become more environmentally sustainable and reducing carbon-based energy sources for food production. Most citizens remain addicted to a form of 'affluenza', including unlimited wealth, affluence and expecting freely available cheap food, while embracing a more destructive form of conspicuous consumption that certainly does not stack up against the growing need for environmental sustainability (Jamieson 2012). Alternatively, calling for a resilient or frugal form of de-growth, including a more holistically conceived, balanced and organic production-consumption model, remains both difficult to visualise, much less promote on film (see Murray and Heumann 2012; Retzinger 2008; Schor 2010; Rust *et al.* 2013; Smaill 2014; Weik von Mossner 2014).

While one suspects the future will lie somewhere between such polarising extremes, Western lifestyles in particular will have to change and this can be sparked by a radical transformation of environmental literacy around public perceptions and behaviour change with regards to food consumption and waste in this instance. Some evidence of this transformation can be garnered from the growth of vegetarianism, organic farming and the slow food movement in particular. Eco-film and media strategies for exploring these issues have developed across a range of academic scholarly readers (see Rust *et al.* 2013; Weik von Mossner 2014; Hansen and Cox 2017), which examine Hollywood fictional narratives and media generally.

This chapter will draw on these scholarly environmental preoccupations, while presenting a close thematic and narrative analysis of three contrasting, even polemical, eco-food documentaries. All are chosen because of their apparent influence on mass audience, as suggested by several studies. Such environmental documentaries can certainly play an important role in reaching a tipping point towards supporting a more sustainable, even organic mode of food consumption. But as will be illustrated in this chapter, some aesthetic and narrative strategies appear more successful than others in imagining this transformation. The documentaries examined all have different strengths, including an incisive use of aesthetic protocols and varying audience appeal. From the popular and successful *Food Inc.*, to the more experimental art-house and poetic style of *Our Daily Bread*, contrasted with *Cowspiracy*, with its first-person narration and immersive style that remains particularly appealing to many new generational citizens, who in turn recognise the environmental need to become vegetarian. All three films are chosen because of their differing formats, from a polemical and preachy style, to a more subtle audience address, alongside more well-established immersive modes of storytelling evident in *Cowspiracy*. Such varying techniques can be called upon as a benchmark for ongoing textual environmental studies, in conjunction with more empirical audience studies that seek to illustrate the effectiveness of audio-visual stimuli. All of which in turn can be used to gauge and promote various levels of environmental literacy and learning, concerning the ubiquitous features and attributes of food.

Literature review

Highlighting the importance of documentary films in speaking to an environmental food agenda, while teasing out how different aesthetic formats of factual filmmaking are co-opted for this project, remains the focus of recent scholarly analysis (see Hughes 2014). Meanwhile, the food industry and for that matter the media and film industry help create their own synergies, while explicitly highlighting the pleasures and inherent benefits of consumption (Baron *et al.* 2014: 81). There are of course lots of powerful fictional films and more recently reality televisual shows and online environmental shorts that are explicitly concerned with food (see Lewis 2008), including for example feature films like: *Chef* (2014), *Julie and Julia* (2009), *Eat Drink Man Woman* (1995), *Chocolat* (2001), *Babette's Feast* (1989). Yet all these fictional narratives, whether consciously or not, ignore agricultural policy and most certainly deny the labour issues that underlie the production of cheap industrialised food. Often purely fictional stories treat food as fashionable or exotic and thereby help to erase the modes of reproduction that underlie our dominant production system, while supporting various forms of commodification that promote the fetishisation of food (Lindenfeld 2010).

Furthermore, big budget Hollywood movies, recalling even environmentally sensitive co-promotions such as *Avatar* (Cameron 2009), surprisingly include partnerships with globally branded companies ranging from McDonalds to Panasonic. *Iron Man* (Favreau 2008) had financial connections with Burger King – as evidenced in the storyline and by using explicit product placement – being the first meal the super-hero has desires on, arriving back home to America. Branding integration with fast food and other consumer products has also been historically central to the *Star Wars* and *Star Trek* tent-pole franchises and is further recalled through the success of Spielberg's *ET: The Extra-terrestrial* (1982), among many other recent examples (see Baron *et al.* 2014: 55).

Stepping outside this bubble of a globalised consumer culture, incisive and critical documentaries are badly needed as a ballast to such sugar-coated fictional exposition of food, which can remind audiences that agriculture is part of nature and culture. This approach further calls attention to a predominant underlying critical message and attitude embedded within much environmental literacy practice. At its most basic, environmental literacy draws clear connections between food production and consumption. Incidentally, publications like *Silent Spring* (Rachel Carson 1962), *The Unsettling of America: Culture and Agriculture* (Wendell Berry 1977), *The End of Nature* (Bill McKibben 1989), *Stolen Harvest* (Vandana Shiva 2000) and other seminal texts continue to influence the underlining drive of environmental documentaries towards expressing the hopes and most importantly the fears and concerns of the sustainable agriculture movement.

As already suggested, mainstream fictional movies appear to mask out concrete representations of industrial modes of agriculture, with the food industry primarily situated as supporting the maximising of profits, by directing attention towards its well packaged end-products, rather than highlighting the often-unseemly means of procurement and factory mode of production. Radical documentaries on the other hand, like those explored in this chapter, illustrate how eco-food solutions need to go back to basics, while dramatising ways in which mainstream and big agriculture remains an intrinsic part of a fractured notion of nature, coupled with all forms of human consumption.[1]

Food security and sovereignty

Unfortunately, in the struggle to enhance profits, food companies tend to obscure, if not actively hide, direct links between the farm gate and the dining table. Yet it has to be continuously recognised that without a food safety net, 80% of the world's population are immediately at risk of hunger, recalling threats of natural disaster, war, etc. (Sheeran 2011). The concept of food sovereignty challenges the dominance of agribusiness and the legitimacy of an unjust global trade system, while alternatively seeking to promote a counter-system of small-scale, localised agriculture, being a fairer solution to hunger, poverty and towards actively addressing climate change. As a broad structuring narrative or master frame, food sovereignty posits clear ethical positions concerning land redistribution, the rights of women, resistance to genetically modified organisms (GMOs) and the defence of local economies (Mann 2014: 7). Furthermore, it is frequently suggested that structural inequalities are primarily responsible for food price hikes, violations of the rights of agricultural workers and the decline in the ability of states to protect their citizens' rights to food (De Schutter 2011 in Mann 2014).

While fictional food films often draw attention to food and culture from a more general perspective and tend to place greater emphasis on eating and drinking as 'an intellectual experience', documentaries more actively highlight concerns about food security and emphasise that 'the act of consuming food is first and foremost a biological necessity' (Baron *et al.* 2014: 180). The three case study documentaries explored in this chapter are specifically chosen because they represent such primal tensions in varying ways and through contrasting registers of engagement.

Slow food movement and class imbalance

Contradictions are keenly apparent in the Slow Food movement for instance, through its predominant class and elitist dimension, which privileges the eating experience beyond primary sustenance for the body. This is affirmed by Carlo Perini, an Italian leftist, who argued that the slow food came to be associated with the pleasure of eating, divorced from any social-political context. A one-time union organiser from the Piedmont region, this founding figure had come to realise that the slow food concept around the right to pleasure in eating had to be broadened to include all those who had the right to such pleasure and thus not become divorced from specific socio-political contexts,[2] echoing similar justice concerns embedded within the environmental movement. The most notable of these cite how various forms of environmentalism appear to be ascribed as a pre-occupation of the rich west wanting simply to conserve, now that they had created great personal wealth at the expense of over-exploiting their communal environment, not to mention their colonial neighbours throughout the world.

The valorisation of food grown ethically and locally is also connected with Aldo Leopold's long-established notion of the 'land ethic', as expanded by Wendell Berry and other environmental thinkers. Consequently, the distinction can be drawn between industrial modes of agriculture, which view the land as a commodity value that simply highlights production-related inputs, as against a local or sustainable food perspective that sees the farmer primarily as a land steward, who fosters a deep agrarian ethos, in which land is closely connected to biological diversity and holistic ecosystems. This philosophy helps to promote a place-based perspective, which is keenly echoed across all three examples, while remaining most critical of the farmer's role, being firmly situated as a key environmental agency and of course a food producer affecting the overall community.[3]

Echoing modern societies apparent desire not to 'waste time' within the process of production/consumption, food sourcing for many and especially those who can least afford 'good food' has become an exercise in acquiring the cheapest cuts, for quick consumption. Furthermore, because of the dominant economic imperative, this reality automatically favours the fast food industry, as demonised in documentaries like *Food Inc.* in particular. Alternatively, the slow food movement and the organic food business generally remains restricted to more environmentally aware citizens, who can comfortably afford such luxuries. Class-poverty remains a key barometer and divide across the environmental ethical debate, with the ever-present communications danger of simply preaching at those who cannot afford the luxury of good food.[4] This danger is constantly highlighted, especially within the left-leaning food documentary format and serves as a cautionary tale, when educating and teaching across various aspects of food production and consumption.

Farming development: from organic to high tech food production

The current global factory farming production and marketing system, according to some reports, is generating a glut of cheap processed meats that is fuelling climate change, while at the same time feeding a pandemic of diet-related ill-health that is costing between 600 and 900 billion dollars a year in healthcare and related costs. All of the documentaries discussed in this chapter, especially *Food Inc.* and *Cowspiracy* draw attention to these worrying trajectories in a very direct and emotive manner.

For instance, if people in developed countries simply ate no more than the recommended levels of meat, more than five million premature deaths could be avoided by 2050, according to Oxford researchers and they further calculate that a worldwide switch to a vegetarian diet would potentially save more than seven million lives. Consequently, progressive environmental food documentaries like *Cowspiracy*, and others explored in this chapter, seek to call attention to such research and actively promote vegetarianism and related forms of sustainable organic food production across all levels of society.

Nonetheless, as recently as 2014, the influential British Chatham House Think Tank found surprisingly low public awareness of the true cost of such dietary choices. However as one might expect, 'consumers with a higher level of awareness were more likely to indicate willingness to reduce their meat and dairy consumption for climate objectives'. Closing what such researchers call this 'awareness gap', is a vital first step towards instigating behavioural food change. Researchers also note the 'sticking paucity of efforts' to help reign in global food consumption and further suggest that governments and environmental groups have been 'reluctant to pursue policies or campaigns to shift consumer behaviour'. The reason seems to be fear of backlash, principally from powerful interest groups, as specifically dramatised in *Cowspiracy*. This fear is acutely evident in America and other Western countries, where food policy is shaped primarily by the agri-industrial lobby.[5]

On the other hand, the ever-expanding food industry states that organic farming, alongside sustainable fishing for that matter, cannot actually feed the world. This remains an overly crude hypothesis and simply presents a binary framing of apparently alternative global production possibilities. There was for instance a time when all farming was organic, with fertiliser primarily made up of compost and organic material. Fields were periodically left fallow (unfarmed) to recover soil moisture and nutrients and crops were historically rotated to prevent nutrient exhaustion and most notably pesticides were non-existent. Farmers, however, in this pre-chemical fertiliser time period remained at the mercy of periodic droughts (despite irrigation) and insect infestations. As populations grew, so did the demand for more food, which in turn lead to more large-scale farming methods being developed and normalised. More recently synthetic pesticides, beginning with DDT came into use during the 1940s and 50s and this practice in turn helped sparked off the contemporary environmental movement.

Furthermore, somewhat counter-intuitively and reminiscent in some ways of climate change denialism, it is implied by some food experts that organic does not automatically equal 'safe'. Africa's and India's crisis in food production for example, described as the global battle with hunger, is largely rooted in a 'soil-health crisis', with new technological methods constantly needed to feed growing populations. Meanwhile, proponents of organic farming argue that this strategy serves to exacerbate the crisis. It should probably be recognised that there are a myriad of variables affecting overall food production levels, as further suggested by the documentaries under discussion.[6]

Organically grown crops on average use 25% less energy than their chemical cousins. By privileging more organic grasses and sustainable methods of production, the major danger of intensive farming affecting climate change at least can partially be kept in check. This remains the underpinning philosophy of several of these alternative food documentaries. Most notably, greater efforts need to be expended in off-setting the fact that factory-based agriculture remains the most water intensive industry on the planet, consuming a staggering 72% of all global freshwater, at a time when the UN says 80% of our water supplies are being over-exploited. The related pillars of food sustainability and environmental security, coupled with global behavioural change around scarce natural resources like water remains a major challenge for all human societies, educationalists and environmental literacy programmes speaking to mass audiences, as these documentaries illustrate.

The Prisoner's Dilemma and other unintended consequences around food consumption

Charles Eisenstein's (2013) characterises what he describes as 'The Age of Separation', as driven by individualism, silo-isation, competition, consumption, ascendency, efficiency, rationalisation, autocracy, concentration of power, managerialism and control. There is no recognition in this worldview of the need to balance these traits with those of trust, redundancy, resilience, diversity, autonomy, localisation, subsidiarity, connectivity, solidarity and cooperation, in order to sustain models which in turn recognise (natural resource) limits, that temper continual and everlasting growth to withstand and avoid overall collapse.

Current world-wide crises that affect energy, water, food, ecological destruction and biodiversity, (mental and physical) health, social and economic failure and climate change, are all merely symptoms of an unsustainable societal construct, which has resulted from a skewed sense of progress that such a pernicious paradigm promotes. Each of these can be solved, according to a 2013 Irish conference on the topic, by what appears at the outset, as an unhealthy mixture of more individualism, more ascendency, more competition, more efficiency, more separation, more rationalisation, more autocratic globalisation and more control.[7]

The most immediate challenge, as characterised by this so-called 'Prisoner's Dilemma', is embracing much needed change and transformation. As global citizens, we would all benefit from collaboration towards the common good. But in an open system with a neo-liberal, free marketplace, promoting weak global politics, cultural distrust and an imperfect communications system does not help the process, with defectors likely to get away with cheating within such a scenario. The only solution to help overcome this, and actively promote doing the right thing for the common good, is to instil robust systems of mutual trust. Yet conventional polities are rarely able to privilege this benevolent philosophy of robust systems of control. Trust relies on insight into past behaviour and cultural preferences, which can be furthered by means of historical investigation, discourse analysis and philosophical disentanglement. Conflict resolution relies on such broad ranging insights, which unfortunately remain in short supply, just at a time when they are badly needed, as evident in our current unstable and fractured global society.

A further stumbling block to successful resolution of such problems revolves around the difficulty of dealing with unintended consequences. When societies do take positive action, they often find that their responses are too limited, or worse still, ending up by opening up a Pandora's Box of other inter-related side-affects and unforeseen problems. Even when recognising and clarifying the right thing to do, many citizens do not actually make the appropriate choice in a timely manner. Like society in general, humans struggle to change harmful habits, repetitive actions, or unhealthy preferences (or path dependencies), even when they clearly recognise the negative consequences of such inaction. As earlier explored in Chapter 3, psychologists and philosophers can certainly provide unique insights into such behaviour patterns and close studies of change behaviour and mapping tipping points from the past, can in turn help render useful insights for the future.[8] All of these problems and historical patterns of action are unfortunately dwarfed by the ongoing difficulties of climate change. Nonetheless, the related pillars of food sustainability and environmental security, coupled with global behavioural change, can at least be highlighted while remaining a major challenge for all human societies as the following three very different documentaries illustrate.

The power of documentary storytelling: *Food Inc.*

Food Inc. effectively draws on the investigative scholarship of Eric Schlosser's book *Fast Food Nation* and Michael Pollan's books *An Omnivore's Dilemma: A Natural History of Four Meals* (2007), as well as *In Defense of Food: An Eater's Manifesto* (2009). As David Denby noted in *The New Yorker*, *Food Inc.* is 'an angry blast of disgust aimed at the American food industry' (95).[9] Yet at the same time it speaks to mainstream audiences across the world, and the film successfully 'steers clear of animal liberation arguments, while pushing for a return to grass-fed, hormone-free meat production' (ibid.: 85).

Interestingly, the powerful food industry's public high-profile campaign against *Food Inc.* remains just the sort of response Hollywood usually aims to avoid. Monsanto, one of the largest agri-food companies in the world still maintains a link in its corporate website to critiques of the film, displaying material designed to discredit arguments set up in *Food Inc.* The website reiterates Monsanto's position that criticism of the company's practices essentially constitutes criticism of America's hard-working farmers. It also emphasises that biotechnology and high-input agriculture are the only real solutions to global demands for food (Baron *et al.* 2014: 78).[10]

Such a globally influential documentary remains a thorny problem for the conventional food industry, because it efficiently articulates that consumer society's drive towards inexpensive food has 'a hidden price tag that takes an enormous toll on our environment, our health, and our society' (Harrington, cited in Baron *et al.* 2014: 85). Rather than focus attention on the ease and pleasure of consumption, as evident in many fictional food films, the documentary alternatively examines food as 'a health issue, an environmental issue [and as] a human rights issue' (Straus, cited in Baron *et al.* 2014: 85). With the documentary's explicitly critical view of the industrial food system stretching from procurement to disposal, it dramatically shows that the rising cost of health care and disaster relief are related to 'the way Americans produce and consume food' (ibid.: 85).

Food Inc. effectively promotes its critical environmental and educational agenda by also using archival audio-visual footage which provides a historical overview of farming and food purchasing changes from World War Two to the present. According to the narrator, a 'deliberate veil' has been constructed between food and its source. In reality, of course in spite of advertising a sanitised form of homogenous factory farming, food is shown as being produced within very large industrial units and is represented as coming from 'enormous assembly lines, where both animals and workers are abused'. With the assistance of contrasting images and montages of these factory farms, interviews and voice-over narration, *Food Inc.* effectively demonstrates how big corporations control all aspects of food production, from seed to final distribution into the supermarket.

Nevertheless, some critics point out that the documentary's assertions are weakened, not only because of an over-reliance on the authority of a narrator, (unlike, at another extreme, *Our Daily Bread*, which is set up as a counterpoint aesthetic to such political documentaries that often suffer from a lack of balance), but also because the documentary argues its points from single examples, with no hint of contestation and that in turn remains unsubstantiated throughout the narrative (Murray and Heumann 2012: 45–48).

Most certainly one could agree that *Food Inc.* 'is meant to be an opening salvo that gets people's attention, not the battle that wins the war' (cited in Murray and Heumann 2012: 48).[11] The high pitched (propagandistic) documentary format further includes individual examples of cattle production and corn production, before offering an alternative. The solution of course demands organically grown food that is presumably meant to also 'help viewers reminisce nostalgically about a pastoral past now out of our grasp' (ibid.: 49).[12]

The opening shot of the documentary for instance frames *us* as an audience flying above a

> bucolic cornfield bathed in soft, warm light. We see exaggerated bright green grass and glowing amber of corn and then cut to images of a friendly red tractor and the quintessential mythic cowboy on a horse, pictured in long shot behind his happily grazing cattle.
>
> (Lindenfeld 2010: 381)

Some critics and environmental media scholars could argue that such aesthetics can in turn serve as a form of greenwashing, by the way it promotes a one-dimensional romantic image of early modes of farming and representations of the landscape. On the voice-over, the narrator highlights that our food system has changed and is been transformed from such mythical pastoral scenes, which are in turn juxtaposed with images of conventional supermarket shelves and clinical factory methods of farming.

The voice-over adds that it will lift the 'curtain that's dropped between us and the food we eat, and how the industry doesn't want us to know the truth about where our food comes from, because if we did, we might not want to eat it' (Lindenfeld 2010: 381). Such direct and overtly constructed scare tactics are common place within such polemical documentaries.

Following numerous interviews, documentary footage and animated sequences, the film divides into chapters that focus on different but related and contentious problems: diabetes and obesity; food borne illness; factory farming; genetic engineering; farm worker protection; pesticides; cloning; environmental impact and the global food crisis etc. There are probably too many areas to cover adequately, but nonetheless this attests to its broad pedagogical approach that in turn feeds into a constant first principle of environmental and educational literacy, namely that 'everything is connected'.

Overall the documentary attempts to engage audiences emotionally (albeit not as effectively as in *Cowspiracy*) by showing deeply personal stories, pointing out the vested interests of power brokers in maintaining the status quo, and unearthing ownership patterns and decisions that determine what eventually lands on our plates. We see food from both the broader perspective of institutions (the power that companies like Monsanto and others hold, recalling the implications of policy decisions) as well as highlighting individual stories, such as highlighting the challenge of poor families struggling with diabetes – juxtaposed with important and pertinent facts – all of which ensures the narrative packs a powerful and emotional punch for mass audiences everywhere.

The finale presents a call to active environmental citizenship. While the soundtrack plays Bruce Springsteen's 'This land is your land', statements flash up on the screen telling us how *we* can rise up and demand systemic change. Audiences are actively encouraged to buy organic, to know what is in our food and to choose seasonal produce that is more sustainable and thereby avoids waste. These are all considered useful pedagogical strategies to actively engage audiences in the message of the text. Viewers are continually reminded, for instance, that the average meal travels 1,500 miles from the farm to the final supermarket and encourages consumers to be more selective in our eating and consumption habits.

However, uncovering how successful such powerful documentaries like *Food Inc.* are in promoting or achieving behavioural change, as well as facilitating active food citizenship, remains an ongoing scholarly challenge. Journalist reviews at least appear to be generally positive with regards to the intrinsic power of such provocative texts. For instance, John Anderson of *Variety* asserts that *Food Inc.* 'does for the supermarket what *Jaws* did for the beach – marches straight into the dark side of cut-throat agri-business, corporatized meat and the greedy manipulations of both genetics and the law' (2009). Meanwhile, environmental film scholars are somewhat less optimistic or encouraging, regarding the potential educational power and influence of such strident documentaries.[13]

More apparently progressive and immersive modes of documentary filmmaking, like that displayed in *Cowspiracy* represent a significant change and a growing tendency within public discourse towards the inclusion of local opinion, not only as a means to depict the realities of individual lives, but also to demonstrate and articulate the voice of the people involved in all aspects of the economy (Hughes 2014: 62). As a result, this strategy can appear more successful in promoting effective new forms of environmental literacy.

Cowspiracy: new generational rhetorical strategies

Produced by celebrity environmental actor Leonardo DiCaprio, *Cowspiracy* also speaks directly to the food industry and effectively uses the conspiracy fictional format, which has a growing appeal, especially for a new generation of audiences.[14] Most notably the documentary uncovers the 'most destructive industry facing the planet today' and 'investigates why the world's leading environmental organizations are too afraid to talk about it'. This shocking, yet often humorous documentary reveals the absolutely devastating environmental impact large-scale factory farming has on our planet and in sowing the seeds of a potentially dangerous environmental and ecological disaster. But I would suggest the documentary presents this shocking story in a more engaging manner at times than *Food Inc.*

As already mentioned, animal agriculture is the leading cause of deforestation, water consumption and pollution, while being responsible for more greenhouse gasses than even the transport industry. This in turn remains a primary driver of rampant destruction, species extinction, habitat loss, topsoil erosion, ocean 'dead zones' etc. Nevertheless, these global worries and environmental fears apparently remain entirely unchallenged across the public sphere and mass media in particular, hence radical documentaries are badly needed.

Cowspiracy suggests animal agriculture is responsible for between 18 and 51% of greenhouse gas emissions and the author and narrator Kip Anderson is shown to have tears in his eyes watching animals being slaughtered. The film's framing at the same time appears more positive than *Food Inc.*, creating an uplifting, action-oriented message, promoting self-efficacy in bringing about change, which left one reviewer feeling somewhat optimistic. 'As well as supporting self-efficacy', which is important for action (Breakwell 1986, cited in Lockwood 2016), the film 'affirmed rather than threatening my sense of self and basic worldview'. Such a strategy has been shown by psychological and behavioural studies to 'create greater openness to risk information' (Kahan and Braman 2006). The documentary further employs a positive affective register, or what Leiserowitz (2006: 48) calls an 'affective heuristic – an orienting mechanism that allows people to navigate quickly and efficiently through a complex, uncertain and sometimes dangerous world, by drawing on positive and negative feelings associated with particular risks' (cited in Lockwood 2016: 3).

The release of a solitary chicken from the slaughter house for example is framed as the pivotal moment in the documentary, where Anderson's personal journey as a consumer meets his investigative journey as a filmmaker.[15] We hear Anderson say that if he cannot bear witness himself to the slaughter of animals for consumption, even in this presumably less environmentally damaged backyard farm [and] that he cannot contribute to any form of animal agricultural practices, for instance. This pivotal moment dramatically breaks up the divide between individual/environment, by collapsing the rationality/affect boundary – which is also arbitrarily defined as anthropogenic – as Anderson emphasises the core message of the film's exposure of violence against both the planet and animals. 'If you care about maintaining the planet's boundaries within safe limits, you must go vegan' (cited in Lockwood 2016: 10).

Commenting on the dangers of speaking out, 'you are putting your neck on the chopping block' talking about it. Consequently, such provocative debates about food become so incendiary, they almost feel like recalling another toxic subject such as nuclear energy, or even, at a stretch, the global challenge of facing up to the illegal drug industry. But, of course, food has always been a huge concern for stakeholders concerned with environmentalism and has most definitely become a touchstone for new generational activism. At the same time, it is difficult to build a responsible consensus involving a set of consistent core values to underpin an environmental educational agenda (Brereton 2016). Strategies involving simply scaring audiences with fears around big business dictating non-environmental policies remains rampant within several provocative food documentaries. Some critics even go so far as to acknowledge their fundamentalist environmental pedigree by adapting this polemical strategy. As also evident within the ubiquitous YouTube industry around the proliferation of food documentaries, such forms of address need to be examined in detail to tease out how educational and effective such online media spin-offs are, with their broad-ranging viewership. Audience research investigation is badly needed to evaluate how successful all forms of mainstream (conventional) mass media outlets are in highlighting sustained pro-environmental attitudinal much less promoting behaviour change.[16]

In any case within environmental textual studies, much work is still needed to tease out the relative aesthetic and narrative importance of more established mainstream documentaries like *Food Inc.* or *Cowspiracy* and unpick their use of celebrities and professional experts in building new audiences, while at the same time ensuring environmental messages are effectively transmitted and received.[17] All of this media exposure appears to signal ongoing transformation and the setting up of possible tipping points for marketing the effectiveness of such audio-visual stimuli and recording their power to stimulate a range of environmental literacy protocols. This discursive process can hopefully help towards supporting transformational food consumption practices through radical behavioural change. At the same time, more cautious scholars believe this is expecting too much from mainstream documentaries and turn to *avant garde* and less preachy food documentaries for such results. Unfortunately, there appears to be less innovative experimentation than might be expected, as we have to go back as far as 2005 and an Austrian film maker for textual evidence of more innovative aesthetic interventions, as suggested by a number of environmental communication scholars.

Our Daily Bread: innovative aesthetics for the future

Reflecting on the ambiguity of contemporary responses to changing landscapes and food production, one of the first great documentary filmmaker to engage in this process is Nikolaus Geyrhalter, who began his documentary career with *Pripyat* (1999) about the no-go-area around Chernobyl and followed this up with a wordless evocative study of industrial farming, titled *Our Daily Bread* from 2005. Most commentators complement the ways in which this documentary implicates the viewer into the worlds' represented, by allowing time to think about the images and by refraining from occupying the space for thought and feeling with unnecessary commentary, dialogue and music, unlike most mainstream documentaries, which bombard viewers with rhetorical and polemical messages. His representation of farming and landscapes in particular, are profoundly changed by technology and the actions of human society and even the contemplation of them suggests the need for material change (Hughes 2014: 14).

Geyrhalter can be compared to his contemporary Austrian filmmaker Erwin Wagenhofer and his film *We Feed the World* (2005), which is further reminiscent of *Food Inc.* For instance, it has a final section which includes an interview with the managing director of Nestle, who robustly defends intensive farming, pointing out the lack of evidence that genetically modified food is harmful to human health and furthermore states that the idea of a public right to water remains an extreme view. Certainly, such right-wing views are very different to environmental activists' belief in the global commons and the need to maintain long term sustainable and balanced land husbandry.

These two very contrasting Austrian documentaries effectively bring this contemporary world and work of agriculture into the public sphere, as a normal and necessary part of the global economy. Both illustrate how the realities of the mass production of food can have very different outcomes. Perhaps the images will be absorbed into the society of the spectacle and accepted as part of the cycle of production and consumption. As part of the pro-filmic production of documentary films, the images have already become commodities after all, but unlike the polemical address of *Food Inc.* or to a lesser extent *Cowspiracy*, reflexive contemplation across the varying environmental issues may hopefully encourage more active and critical engagement as part of an educational and literacy programme.

Our Daily Bread focuses on the latest industrial forms of agriculture production, while contemplation of the landscape creates a sense of cognitive dissonance through the economic conflict set up between the benefits of modern factory farming as against the clearly visible and detrimental effects on landscapes, as well as on working practice. Such artistic documentaries raise awareness through spectacle rather than verbal rhetoric and hyperbole, which in turn serves to close down active engagement with the complexity of the debates presented. Instead such narrative tropes help set up the pervasive demand to revise beliefs about the resilience of the natural world to withstand the impacts of human civilisation and its intensive agricultural methods.

Scholars appear to suggest that *Our Daily Bread* comes closest to capturing the three-dimensional 'truth' of food production and consumption, offering fragmented observations that closely replicate the segmented process of industrial food production, effectively revealing its consequences for human and non-human nature, because the intermediary veil of direct cinema has been lifted. 'It provides a nostalgic view of traditional farming methods as a contrast to industrial methods currently employed in Europe', without background sounds and voices to support its visual rhetoric – the *avant garde* rhetorical documentary 'relies exclusively on visual rhetoric' (Murray and Heumann 2012: 44).

The documentary nonetheless makes both the reality and the romantic myth of the pastoral transparent, without diluting the powerful visible rhetoric on display. Manohla Dargis further highlights this aspect in her *New York Times* review (24/11/2006) and demonstrates how the wordless documentary uses 'cool deliberation, showing rather than telling through the long tracking shots'. Meanwhile, Leslie Felperin of *Variety* (23 December 2005) describes the visual rhetoric of *Our Daily Bread* with its 'precisely composed lensing and painstaking sound [which] create moments of sublime beauty, even when showing the production-line slaughter of animals'. In this mechanised world, which *Food Inc.* also captures, but here not just the animals, the human workers are also portrayed as alienated, unable to escape the sounds of machinery that accompanies even their lunch breaks.[18] But of course as with all experimental and *avant garde* film making, one has to question if mainstream audiences are actually viewing such (sometimes difficult) semiosis, or is it only speaking to the converted and those who already have a high degree of environmental literacy.

The opening sequence reminds me of the highly stylised film drama *Hunger* (McQueen 2008) which depicted the political prisoner and martyred hunger striker Bobby Sands from the political conflicts of Northern Ireland. Both films spend significant screen time observing a worker cleaning the ground and sweeping impure liquids away – alternatively human and animal. But through the *mise-en-scene* of *Our Daily Bread*, audiences observe displays of rows upon rows of dead animal carcasses in a huge warehouse. This sense of routine time passing, while evoking large industrial spaces, echo across all forms of factory farming, from chickens and cattle to even hot-housed vegetable production. At one stage, audiences observe a solitary worker donning full body protection including breathing apparatus, as he proceeds to spray chemicals across large swaths of glass-housed plants. The environmental worries relayed in Rachel Carson's 1960s classic *Silent Spring* come to mind. A follow-on sequence counterpoints this scenario, with a shirtless man picking the bountiful fruits from the plants for human consumption.

The repetitive nature of such factory work and the isolation of workers is dramatised, as audiences constantly witness them consuming food on their own or smoking a cigarette during a solitary work-break. Otherwise, the workers seem to have nothing else to relieve their boredom. For the general viewer, however, audiences instinctively expect background music and a controlling voice-over, or alternatively at least dramatic interviews, to frame a consistent meaning and help bridge the interpretation process for the reader. While film scholars seeking new generic aesthetics and developing a more nuanced form of media literacy and engagement admire the openness of the narrative and the prosumer's capability to make up their own mind and interpretation, the images spill into each other without any driving or coherent storyline. By all accounts the albeit opaque message incorporates a pervasive critique of the Fordist model of mass production, while highlighting an overall lack of concern for any long-term resolution with regards to sustainable environmental issues. All that matters in this form of agribusiness is the bottom line of mass production and constantly enforcing reduced costs. Consequently, a more reflexive aesthetic model of film making is privileged, rather than the shrill montage exposition and preaching of *Food Inc.* or even recalling the more inclusive immersive storytelling model evident in *Cowspiracy.* Nonetheless, this chapter argues that because of audience interests and varying levels of literacy, *both* polarising forms of documentary expression and more besides are necessary to help promote environmental food literacy and most importantly encourage behavioural change across a broad range of audiences, while tapping into inter-generational taste cultures.

Concluding remarks

The toxic materiality of the eco-documentary, according to Helen Hughes in her provocative 2014 *Green Documentary* monograph, is a matter of a complex network of social and material effects, involving not only the immediate material of the DVD or film strip, but also the design and mass manufacture of technology, travel and transportation, land use and most importantly accessibility. As already implied, *we* need all types of food documentaries, from the shrill polemical varieties to the more immersive first-person narration, alongside the more nuanced *avant garde* format and much else besides, to help in the struggle towards activating environmental engagement and help to develop new forms of food literacy.

Environmental awareness, as defined by Kollmuss and Agyeman, includes simply 'knowing of the impact of human behaviour on the environment' (2002: 253), and serves as the goal of many different kinds of activity concerned with education about human impact on the environment, including food consumption.[19] However, according to such research, one of the earliest findings of environmental behaviour research in the 1970s was that knowledge and awareness is not sufficient in itself to lead to radical changes in behaviour. Yet most environmental non-governmental organisations (NGOs) – and one could add food sustainability documentary approaches alluded to above – 'still base their communication campaigns and strategies on the simplistic assumption that more knowledge will lead to more enlightened behaviour' (Hughes 2014: 12). Through ongoing psychological and behavioural studies in the media, this has been shown however not to be the case. There is no magic bullet solution towards communicating environmental literacy, nonetheless there are relative merits in choosing one particular documentary aesthetic over another.

For instance, Greg Mitman talks of a contemporary 'green wave' of film and television, facilitated by a popular penchant for 'eco-chic' that is underpinned not only by commercial imperatives, but also by counter-balancing ethical and environmental concerns. Mitman notes that of the 631 million in gross revenues earned by 275 documentaries released between 2002 and 2006, 163 million came from eight wildlife and natural documentaries (216). In trying to speak to large uncommitted audiences, environmental food documentaries can learn a lot from the broad-ranging pleasures and aesthetic protocols set up within more mainstream nature documentaries, alongside other related fictionalised environmental narrative formats (see Brereton 2016).

All the while threats to the food supply chain represent a cogent and visceral concern for us all – especially as food speaks so vividly to both short and long-term human needs and related manifestations of global environmental transformation. The search for food amidst a world of scarcity remains a frequent motif for instance within numerous post-apocalyptic science fiction fantasies, alongside being directly highlighted inside the documentaries discussed in this chapter.[20]

By any measure, food remains one of the most important attributes, triggers and barometers of environmental ethics and literacy, as consumers are directly implicated in maintaining food prices as low as possible, to help maximise competitive productivity and even some argue to maintain long-term sustainability. Meanwhile deep tensions between factory farming, as opposed to more organic forms of (slow) food production, speak directly to a broad range of environmental and ethical tension and agendas, all of which has huge implications for day-to-day living for humans across the world. With expediential growing populations and increasing demands for corrective economical models for increased food production, there are grave dangers involved with un-regulated systems of food production, distribution and consumption.

Such developments for instance, do not consider the long-term common good of society in general, much less recognise further concerns around for example the Precautionary Principle feeding into issues of food quality, health and security. As evident since Rachel Carson's seminal and prototypical literary text *Silent Spring*, once a direct health concern is raised – making a clear connection between pesticide and the human food chain for instance – then such concerns have to be urgently addressed. More recently, worries over the safety of food has reached a tipping point and has spiralled into ongoing risk-debates around GM-food and what is deemed as acceptable practice, much less taking into account what is natural or normative with regards to mass produced factory food production, security and consumption.

Appealing to public interests concerning the quality of food can serve as a useful bridge between cultures, including those that appear to place a more holistic value on good quality organic food and those countries like America that seems to privilege the economic bottom line. Related forms of environmental food literacy projected onto the inter-related spectre of climate change remains an ongoing big picture concern for all forms of media representations, as explored by environmental communicators. Certainly, the lack of environmental literacy and clear discursive practices of communication around food, stands out as a pervasive danger and risk (Lindenfeld 2010: 383). Documentaries continue to have an important part to play in the targeted communications agenda around exposing environmental issues to mass audiences. However, in striving to find new ways to address audiences, documentary formats alongside fictional narratives, need to provoke more open engagement and adopt new aesthetic strategies that hopefully can succeed by cross-connecting with other forms of new media, while speaking to evolving food debates and related environmental concerns. The three documentaries explored in this chapter have been shown to be exemplary in teaching environmental food literacy. But much more creative imaginaries are needed in the future to connect with new generational audiences.

Notes

1 *The Corporation* is one of the most economically successful documentaries to detail problems in the food industry, yet the film made only 3 m at the box office. *Flow* only raised 142,569 gross in the US market, *King Korn* 105,422 dollars etc. Available online at: www.BoxOfficeMojo.com

See also top 20 Documentaries and box office receipts up to 2012 which include: *March of the Penguins* (2005) at number 2 with 77 m, *Earth* (2009) at number 5 with 32 m and *An Inconvenient Truth* number 9 with 24 m … down to number 17 with *Super-Size Me* (2004) at 12 m (Baron *et al.* 2014: 185).

2 This concern about equity and essentially speaking to a class bias, led him and others to add 'Fair' to the slow food slogans of 'good' and 'clean', emphasising that the food should be grown sustainably.

3 By all these measures, a Farmer's Market meet all the criteria: food grown locally; tastes good; fair food; purveyed in a consistent manner with the food justice framework; and clean food as it is grown sustainably.

4 Nonetheless, food not 'only shapes our bodies, but it structures our lives, fashion daily rituals and helps to mark significant rites of passage. Food connects us to others –

both directly, through shared meals, and culturally, through shared tastes' (see Retzinger 2008: 370).

5 For instance, high-tech alternatives to meat, such as the approach of 'Beyond Meat' by the US food company backed by Bill Gates are emerging as possible solutions to the problem. This is activated by using plant-based ingredients to closely mimic the taste, texture and smells that make meat so alluring, while apparently setting out to trim the health effects and global environmental havoc that our ancient carnal predilections are now wrecking.

6 Experts, like Ed Hammer and Mark Anslow remain more positive regarding the overall benefits of organic farming, while asserting in *The Ecologist* from 3 January 2008 that *we* all need to cut down on meat consumption in particular to help normalise the average consumption of one and a half pounds per person per week (43–46).

7 Available online at: www.ucc.ie/en/media/conferences/sustainabilityinsociety/OrientationChapterTransdisciplConversationsConf2013.pdf (Introduction to 2013 conference on Sustainability in Cork University 5–6 September: 'New Paradigm Thinking: Alternatives and Visions Transcending the Disciplines.) Of course, we would question such assumptions and hypotheses.

8 See www.mdpi.com/journal/humanities Humanities for the Environment.

9 While, for example, *Supersize Me* (2004) took a massive box office revenue of 11.5 m, *Food Inc.* by 2009 had overtaken it by 4.5 million. Unfortunately, the other two documentaries explored in this chapter did not reach the top 100 earners in the box office charts but are available on Netflix and online for new genre audiences to find. In the IMDb best food documentaries list *Our Daily Bread* is rated number 11, with *Cowspiracy* at 71 (June 2018).

10 For an overview of Monsanto's latest position regarding the documentary see its website (accessed 3 June 2018) where it justifies its patenting of GMO seeds etc for the betterment of farming in general. https://monsanto.com/company/media/statements/food-inc-documentary

11 According to Andrew O'Hehir (2009) from the online magazine/website Salon.com, 'the food-activism movement in 2009 is roughly where the environment movement was in 1970'.

12 Nonetheless, *Food Inc.* adapts an anthropocentric perspective on food that relegates environmental issues into peripheral status, which of course is the norm in the USA and elsewhere. The film's style by all accounts remains highly effective due to its high production values and its narrative structure is engaging and entertaining.

13 As already suggested, Murray and Heumann suggest that while *Food Inc.* provides a plethora of information about industrialised farming and argues from a clear position, its message is weakened by 'both its nostalgic vision and by the rhetorical strategies the film-maker choose to employ' (2012: 49).

14 See *The Sustainability Secret* produced by Kip Andersen and Keegan Kuhn: view www.cowspiracy.com – short promo to examine its potency.

15 In *The Transmission of Affect*, Brennan (2003: 14) points out there is no secure distinction between the 'individual' and the 'environment' which many food documentaries find it difficult to tease out.

16 Note many of which feature a number of specialists and food celebrities (including Richard Oppenlander, Michael Pollan, Will Tuttle and Howard Lyman).

17 Anecdotally, colleagues have talked how their children have been dramatically affected by such documentaries in recognising the long-term environmental dangers of industrial food production and of them being willing to radically change their life styles and food consumption patterns as a result.

18 Death is clean in the factory farm, the film shows us, and workers are cold and distant, consuming food they prepare and seemingly oblivious to the unnatural state of the food production process (Murray and Heumann 2012: 56). See almost 10,000 views on YouTube. Available online at: www.youtube.com/watch?v=ddxo7QSiqfs.

19 A further goal of environmental education is the promotion of pro-environmental behaviour that as Kollmuss and Agyeman put it in their survey of approaches to the subject, means 'behaviour that consciously seeks to minimise the negative impact of one's actions on the natural and built world (e.g. minimise resource and energy consumption, use of non-toxic substances, reduce waste production)' (2002: 240).

20 When hunger takes a literal rather than a metaphorical form, it propels actions that serve to define what it is to be human – or to be inhuman. Food and water scarcity lead both to brutality and kindness in science fiction films such as *The Omega Man* (1971), *Mad Max 2* (1982) etc.

References

Anderson, J. 'Food Inc: With a Constituency Limited to Anyone who Eats, Food Inc. is a Civilised Horror Movie for the Socially Conscious, the Nutritionally Curious and the Hungry' *Variety*. 11 September 2008.

Baron, C., Carson, D. and Bernard, M. *Appetites and Anxieties: Food, Film and the Politics of Representation*. Detroit MI: Wayne State University Press. 2014.

Brennan, T. *The Transmission of Affect*. New York: Cornell University Press. 2003.

Brereton, P. *Environmental Ethics and Film*. London: Routledge. 2016.

Dargis, M. 'What's for Dinner: You Don't Want to Know' *New York Times Review*. 24 November 2006.

Denby, D. 'Anxiety Tests: The Hurt Locker and Food Inc' *The New Yorker Review*. 29 June 2009.

Eisenstein, Charles. *The Ascent of Humanity: Civilization and the Human Sense of Self*. Berkeley CA: North Atlantic Books, Evolver Edition. 2013.

Felprin, L. Review of 'Our Daily Bread' *Variety*. 23 December 2005.

Hammer, E. and Anslow, M. '10 Reasons Why Organic Farming can Feed the World' *The Ecologist*, 43(6). 2008. Available online at: https://theecologist.org/2008/mar/01/10-reasons-why-organic-can-feed-world.

Hansen, A. and Cox, R. (eds) *The Routledge Handbook of Environment and Communication*. London: Routledge. 2017.

Hughes, H. *Green Documentary: Environmental Documentary in the 21st Century*. Bristol UK: Intellect Books. 2014.

Jamieson, D. *Ethics and the Environment: An Introduction*. Cambridge UK: Cambridge University Press. 2012.

Kahan, D. and Braman, D. 'Cultural Cognition and Public Policy' *Yale Law and Policy Review*, 24: 147–171. 2006.

Kollmuss, A. and Agyeman, J. 'Mind the Gap: Why Do People Act Environmentally and What Are the Barriers to Pro-environmental Behaviour' *Environmental Education Research*, 83: 239–260. 2002.

Leiserowitz, A. 'Climate Change Risk Perception and Policy Preferences: The Role of Affect, Imagery and Values' *Climate Change*, 77: 45–72. 2006.

Lewis, T. 'Transforming Citizens? Green Politics and Ethical Consumption on Lifestyle Television' *Continuum: Journal of Media and Cultural Studies*, 22(2): 227–240. 2008.

Lindenfeld, L. 'Can Documentary Films like *Food Inc.* Achieve their Promise?' *Environmental Communication*, 4: 378–386. 2010.

Lockwood, A. 'Graphs of Grief and Other Green Feelings: The Uses of Affect in the Study of Environmental Communication' *Environmental Communication*. 2016.

Mann, A. *Global Activism in Food Politics: Power Shift*. Basingstoke UK: Palgrave. 2014.
Mitman, G. *Reel Nature: America's Romance with Wild Life On Film*. Seattle WA: University of Washington Press. 2009.
Murray, R. and Heumann, J. 'Contemporary Eco-Food Film: The Documentary Tradition' *Studies in Documentary Film*, 6(1): 43–59. 2012.
O'Hehir, A. 'Best Movies of 2009' *Salon.com*. 2009. Available online at: www.salon.com/2009/12/28/aoh_2009/.
Retzinger, J. P. 'Speculative visions of Imaginary Meals: Food and the Environment in (Post-apocalyptic) Science Fiction Films' *Cultural Studies* 22. London: Routledge. 2008.
Rust, S., Monani, S. and Cubitt, S. (eds) *Econcinema Theory and Practice*. London: Routledge. 2013.
Schor, J. 'The Principle of Plenitude' *Minding Nature*, 3(2). 2010. Available online at: www.humansandnature.org/the-principles-of-plenitude.
Sheeran, J. 'Wages the Ultimate Food Fight' *The Globe and Mail*. 2011. Available online at: www.theglobeandmail.com/report-on-business/careers/careers-leadership/josette-sheeran-wages-the-ultimate-food-fight/article598360/.
Smaill, B. 'Documentary Film and Animal Modernity: An Analysis of Raw Herring and Sweetgrass' *Australian Humanities Review*, 57: 56–75. 2014.
Smaill, B. 'New Food Documentaries: Animals, Identification and the Citizen Consumer' *Film Criticism*, 39(2): 79–102. 2014.
Weik von Mossner, A. (ed.) *Moving Environments: Affect, Emotion, Ecology and Film*. Waterloo ON: Wilfrid Laurier University Press. 2014.

5 Eco-documentaries

Old problems, new aesthetic opportunities

Overview

Greta Gaard (2009: 327–330) raises 'three important questions' about the capacity of children's (and young people's) exposure to environmental literature to increase ecological literacy, which can equally be applied to all forms of audio-visual media: 'First, how does the text address the ontological question, "who am I"', 'Second, how does the narrative define the eco-justice problem?', 'Third, what kind of agency does the text recognise in nature?'

Frequently, scholars have suggested that climate change attitudes and belief amongst undergraduate groups of students in particular have been historically under-sampled or examined. It is suggested that this group (while still exhibiting the same partisan divide, evident in other age brackets) may be slightly more accepting of the reality of climate change than older members of society (Burkholder *et al.* 2017: 4). This hypothesis has fed into this study's investigation of an apparent generational divide, recalling imbalances with the growing consumption of new media in particular. Certainly, extensive empirical research is required to tease out such assumptions and in helping to understand and target improvements to general environmental literacy across new-generational audiences.

The previous chapter on food documentaries presented the traditional model of forthright, even didactic engagement with such prescient environmental issues and certainly did not hold back. In this new age of prolific online media and in recognising a lack of consensus around a Public Service Broadcasting frame for media education, with some exceptions general audiences appear less predisposed to being preached at. 'Hitting the sweet spot' between the extremes of pontificating and alternatively being overly complicit in supporting a 'business as usual' commercial mentality, captures an ongoing tension surrounding contemporary environmental documentaries.[1]

While drawing on a number of well-established behavioural theories, including for instance nudging and appealing to PR and celebrity endorsement, this chapter will examine two contrasting environmental documentaries. This chapter will highlight concerns over being too strident, as evident with *An Inconvenient Sequel*, compared with adapting a more light-hearted, magazine style and clearly celebrity-driven approach to environmental concerns, as illustrated by an analysis of *James May's Big Ideas*. Surprisingly, the latter format can be shown to be more environmentally enticing for mass audiences. As Tania Lewis asserts, such otherwise ephemeral television shows 'represents an important space where ethical and moral questions concerning the relationship between the privatized citizen and the broader community are played out' (2008: 227).

Many environmental scholars have become extremely worried at the lack of recognition of the extreme dangers facing our planet. Witness for instance the short-lived effects on policy around droughts or even freak storms such as Katrina (2005), Sandy (2012), Haiyan (2013) and most recently Harvey (2017), Emma (2018) or the Carr fires in California (2018). These external catastrophic events apparently are not enough to spark a powerful trigger to help bring about a cultural and political shift. Scholars like the Norwegian psychologist Per Espen Stoknes (2015) suggest that what is needed is the 'work of a cultural movement similar to the ones that dismantled apartheid, abolished slavery, or took on nuclear arms'. Certainly, this ongoing dilemma poses a major challenge for the Humanities, coupled with Social Science disciplines and feeds off a range of psychological and behaviour barriers addressed in Chapter 3. Hopefully new forms of mass media can assist in calling attention to the urgent need to help solve a broad range of these inter-connecting environmental issues, coalescing around the spectre of climate change.

Stoknes proceeds to signal five major barriers to effective climate change communication, which remains especially prescient for all forms of new media literacy and echoes the broad range of psychological barriers to effective communication, as suggested in previous chapters:

- Distance – where stories remain remote for the majority of us
- Doom – which can only be addressed by loss, cost and sacrifice and avoiding the topic
- Dissonance – where style of living, including for instance flying and driving, conflicts with environmental demands, consequently the message backfires
- Denial – when we negate, ignore, or otherwise avoid acknowledging the unsettling facts about climate change. We therefore find refuge from fear and guilt
- iDentity – where we filter news through our professional and cultural identity.[2]

Citizens constantly look for information that confirms their existing values and filter out what challenges them. New media social networks, for instance, are especially effective in keeping its users within a range of bubbles. Paradoxically this pulls against the very notion of the Internet and new media in general, as providing an open window on the world, compared with more conventional media or documentaries for that matter. If citizens who hold conservative values, for instance, hear from a liberal that the climate is changing, they are less likely to believe this message. Such evaluations can certainly be used to frame and critique the two very contrasting documentaries under discussion in this chapter, emanating from opposite poles across the ideological spectrum. Nonetheless,

I suggest, in the end, both of them can succeed in their own way, by reducing the distance of the story and their intended audience. Like all effective documentaries and mass media generally, the documentary format efficiently collapses time and space and helps to bring the viewer directly into the storyline, no matter who they are, or the values they might hold dear.

An Inconvenient Sequel (AIS), as one might expect, is unfortunately more prone to presenting doom scenarios, compared with the magazine format appeal to technological solutions for energy provision in *James May's Big Ideas* (JMBI). Certainly, dissonance remains the trump card for JMBI, with its youth audience probably aware that it is impossible to square the excessive celebration of car (oil-based) culture with any notion of environmental rectitude and resistance. Alternatively, one could argue, AIS overly pontificates to its audience from a position of scientific certainty. Furthermore, JMBI has a clear lead in the denial stakes, while also embracing iDentity modalities as outlined above. Yet such a magazine programme would, at the very outset be perceived with good cause by most environmental communicators, as being both reckless and somewhat loose in its generic address, towards adequately addressing the struggle towards promoting environmental truth. Nonetheless, this chapter will strive to counter this evaluation, by showing that in spite of so many caveats, such a light-weight environmental format remains important in communicating environmental issues and concerns to its broad-based younger audience, compared with the more earnest and committed AIS model. Such a comparison will help to illustrate that both polar opposites have much to offer in helping to develop new generic modes of environmental engagement into the future.

As has long been implied by environmental communication scholars, fictional strategies of fear and loss don't necessarily appeal to audiences. Consequently, new models of proactive engagement and the creation of robust environmental and critical literacies are needed, which in turn are more conductive to action and suggesting behaviour change solutions. Most usefully and pragmatically, Stoknes believes we should at least begin to talk about climate in terms of 'insurance, health, security, preparedness, and most of all, opportunity'. As argued across much ecocinema scholarship, the media need to tell more powerful and seductive stories, drawing on the benefits of a creative imagery that breaks our entrenched social and ideological bubbles, while provoking strong emotions associated with such investment. 'To be truly radical today is to make hope possible, not despair convincing', as Raymond Williams once said. While still believing there is a place for dramatic apocalyptic narratives to help warn against future disaster, creative learning models of environmental literacy are becoming more pertinent and need to go one step further in actively speaking to more positive attributes around the 'good life' and supporting general human well-being. All of which serves to expose the unique attributes of humans thriving within a benevolent and sustainable environment, while actively promoting new forms of green growth and supporting long-term stewardship of the earth.

As already noted, as a very crude and simplistic rule-of-thumb model, while recognising the danger of succumbing to a utopian ideal, most narratologists instinctively agree that when telling (audio-visual) stories, they should be:

- Personal and concrete
- Vivid and extraordinary
- Show, don't tell
- Wherever possible, make them humorous and witty, with a strong plot and engaging drama.

Such a simple road map of basic rules of good fictional and factual documentary-style storytelling formats certainly corresponds with the ongoing successful model embraced by big budget Hollywood features and more recently has been adopted across all commercial media platforms and generic formats. Of course, using engaging artistic strategies are not enough in themselves to ensure societal and behavioural change. Nevertheless, this chapter illustrates how more can be learned from adopting business and marketing protocols, which have grappling with such difficulties over a long time – recalling echoes of debates explored in the greenwashing case study cited in Chapter 3. It may seem ideologically suspect to envisage new forms of critical learning being garnered from commercially-driven business scholarship that ostensibly support a consumption-model of capitalist growth and which in turn remains anathema to sustainable environmental progress. Nonetheless, in the real world, environmental communication scholars need to uncover hope and embrace pragmatic responses and the possibility of effective solutions wherever they can find them. For example, the highly-regarded (business) theory of nudging is as good a place to start as any, promoting this all-encompassing agenda of active cross-disciplinary environmental learning and productive communication.

Nudging audiences to support environmental change and transformation

The majority of audiences appear less concerned with worries around climate change and at least unconsciously privilege a 'business as usual' model of economic GDP-targeted development. Most citizens almost implicitly appear committed to the 'idea of homo economicus – or economic man – the notion that each of us thinks and chooses unfailingly well, and thus fits within the textbook picture of human beings offered by economists' (Thaler and Sunstein 2009: 7). Environmental scholars alternatively strive to question and counteract such one-dimensional logic and what they consider faulty economic rationality, while addressing the multi-layered difficulties embedded across climate change issues. At a conservative level, one can assess that, for instance, so-called light ecological thinking tries to co-opt such pervasive thinking in an attempt to begin to make modest changes in the system. In particular, as embedded within cross-disciplinary marketing and finance instruments, nudging can be fruitfully applied and filtered throughout the general environmental lexicon.

A nudge is any aspect of the 'choice architecture that alters people's behaviour in a predictable way without forbidding any options or significantly changing their economic incentives'. To count as a mere nudge, the intervention must be easy and relatively cheap to carry out. Nudges are not mandates. 'Putting the fruit (in a shop) at eye level counts as a nudge. Banning junk food does not.' Recalling the cautionary literature on the dangers of Greenwashing and at the same time encouraging companies to adopt and buy into their ever expanding 'Corporate Social Responsibility' ethos, appears an obvious strategy for quick if easy gains across large swathes of productive mainstream society. By all accounts 'private companies that want to make money and to do good, can even benefit from environmental nudges, helping to reduce air pollution (and the emissions of greenhouse gases)' (Thaler and Sunstein 2009: 6). Similarly, the commercial mass media industry, which remains predominantly profit-driven and focused on the maximisation of audiences to support advertising revenue, can at the same time uncover reputational value and become almost inspired, while actually supporting a green corporate social responsibility ethos.[3]

From explicit environmental programmes transmitted on mainstream television, to green-washed advertisements, to highlighting celebrities who use their power and influence to address a range of environmental issues (see Miller 2017), all of these approaches and media formats can be used to initiate a broad range of environmental nudges. From encouraging a love and respect of nature (biophilia), alongside appealing to an ethical engagement with a number of difficult environmental issues; these mainstream strategies posit a series of helpful nudges to actively encourage a constructive engagement with environmentalism. This approach is highlighted within the complicit environmental aesthetic format of *James May's Bright Ideas*. Consequently, it can be usefully co-opted to help call attention to the importance of environmental concerns, making it more palatable for a broader audience to engender attitudinal and behaviour change. Being pragmatic and recognising the scale of the problem, it is essential for media communications to address all strata of society across the world – not just those already sold on the environmental agenda and its sustainable philosophy as in *An Inconvenient Sequel* – to face up to their primary responsibility with regards to climate change.

Greenwashed documentaries: from *Top Gear* to *James May's Big Ideas*

To forcefully address environmental engagement and understand audience antagonism towards overtly conceived environmental documentaries, we need to examine how and why more mainstream anti-environmental narratives have become so popular and globally successful of late. For instance, one could cite (BBC's) *Top Gear*, as one of the most popular ever cult television magazine series aired in the UK beginning in 1977. The initial outing was unsuccessful until the 'right formula' was adopted from 2002 to 2015, when the series managed to become a 'lightning rod for a wide range of political issues, including around climate change and anti-environmentalism' (Drake and Smith 2016: 701). Essentially the series recalls guilty pleasures around celebrities indulging in driving fast cars.[4] According to BBC Worldwide, by the end of 2014 the programme's YouTube channel had over four million subscribers and over 800 million video views. By any measure, such a PR-driven media phenomenon remains the antithesis of stoical environmental sustainability and the necessary (re)presentation of alternative low-carbon energy futures.[5] Drake and Smith paint a useful critique around how the series actively targeted the environmentalist lobby, which seeks to 'cut carbon emissions (and thus implicitly curtail the unnecessary use of vehicles that burn petrol or diesel)' (2016: 690).

Back in 2008, one of the original presenters James May – who probably is more amenable to the science of environmentalism than his fellow frontman Jeremy Clarkson[6] – has re-purposed his technical skills to present a science-based spin-off celebrity magazine format programme, not surprisingly titled *James May's Big Ideas*, which is still finding new audiences, being freely available on YouTube.[7] An early section of the final episode on energy begins by parodying, even appearing somewhat bemused at, the very idea of society moving towards a 'low carbon energy future'.[8] Incidentally this theme constantly preoccupied *Top Gear*, whose *raison d'etre* embodies the very antithesis of environmental frugality, much less contemplating any move towards de-growth for that matter. The well-known presenter starts the final episode in the 2008 spin-off series by affirming how *we* simply want more energy to keep everything going, not accepting in any way that the 'party' has to stop. Of course, the pervasive use of 'we' across such discursive rhetoric serves to ideologically 'hail the audience' with the programme's prescribed and uncontested 'common sense' belief system. No doubt however this position and attitude also remains the popular mindset and belief system of mass audience engagement, if often not so explicitly recognised in the educational academy; namely, that *we* the general public want to have it every way. Mass consumer society essentially want to maintain their high carbon and exciting fast-paced lifestyle, while at the same time probably wishing to embrace wholesome environmental values. These incompatible positions correspond with protecting what can be dismissively characterised as a 'business as usual' point-of-view, while also wanting to save the planet. One could posit such a set of incompatible positions is prevalent across so many (non-environmental) stakeholders and communities.

Essentially, looking for quick fix technological solutions to any potential energy shortage, while avoiding or even alluding to the radical proposition that *we* have to be 'weaned off fossil fuels' in the first instance, is conveniently ignored. The series whole-heartedly accepts the status quo and simply posits the need to discover *more* sources of energy and quickly, to help secure future stability. For instance, uncovering the hidden benefits of the sun is one idea, which audiences are reliably informed is 92 million miles away and has great power to be harvested to satisfy the needs of humanity's ever-expanding energy demands.

James May begins his techno-fix series of solutions to climate change by first of all demonstrating a student-made solar powered car and showing how such a vehicle can point the way, but only when the actual applied technology has greatly improved. Next, he visits a solar plant in the south of Spain near Seville where there is a massive tower (reminiscent of the *Lord of the Rings* franchise) to focus all the solar panels, while reflecting the energy of the sun back into it. This technological innovation is simply using 'fancy plumbing', May seductively explains. This can in turn be transformed into producing efficient and sustainable energy into the future.[9]

Later, as one might expect, the presenter suggests that the Sahara Desert and solar plants could supply the whole of Europe with enough energy for its needs, but he quickly concedes that there are some flaws with this proposition, while not even mentioning the unsustainable cost of distribution and transferring energy all across the continent. In the end, May calls upon an even more contentious 'blue skies' solution and suggests that *we* need to think beyond the surface of the earth and go where the sun is always shining, out into deep space. Not any more preposterous, one must admit, than the storylines of so many recent science fiction fantasies from *Elysium* (2013) to *Interstellar* (2014) or *Passengers* (as discussed in the following chapter). NASA appears to be offering a '2-million-dollar reward for the team that develops the best prototype for a Space Elevator'. Is this a fanciful idea or what – using human ingenuity to escape the 'problems' of climate, by actively supplying unlimited electrical energy for a future techno-fix nirvana? Unlike so many (science) fictional narratives, however, there are no worries raised about taking into account security checks and balances, including calling to mind the precautionary principle, or dealing with possible unintended consequences, much less recognising the huge resource costs to get such a massive project to completion, not to mention the inevitable time delay involved, with no security or certainty that such investment will deliver productive results in the long run.

Like *Top Gear* and its clones that seek out the latest gadget or the thrill of even faster speed, this populist-driven, pro-consumption magazine-style documentary functions as an unambiguous celebration of technological determinist and geo-engineering solutions to our climate change energy dilemma. Few would even expect such a complicit programme to acknowledge, much less accept, the more difficult ideological and political position that humans as a species have to radically change and adopt their behaviour and mode of living, if they are to survive in the long term on planet earth. Furthermore, pushing the analogy further, the logical development of such a *Top Gear* cloned mind-set is visualised through the excesses of a *Mad Max* franchise type storyline. This involves dramatising a dystopic future world with exciting but lethal skirmishes over scarce oil resources, while portraying a generally dysfunctional and atomistic society struggling to maintain its insatiable energy needs, to help keep the show on the road. Such representations feed off the continuous legitimation of a de-regulated market and in turn a chaotic drive to the bottom.[10]

All the while, audiences are encouraged to wonder, can human ingenuity actually produce a kind of 'non-toxic' form of carbon excess. Imagine if James May's pro-carbon point-of-view was alternatively flipped on its side and supported by actively reading against the grain to posit a selection of oppositional visions? With some careful re-modelling, such creative imaginaries might furthermore even help adjust mainstream aesthetic sensibilities away from wallowing in oil-based fantasies. Such oscillating positions might be parodied by this magazine format, using for example Formula One racing, which is constantly reified as the 'Holy Grail' of driving experience within *Top Gear*. Might such dangerous (anti-environmental) representations be somehow re-appropriated and re-purposed to alternatively calling attention to more productive low energy sources and creative narratives? What kind of subtle or more substantial shift in formal aesthetics and perception would this require, while helping audiences to actively perceive the future beyond our carbon-saturated imagery? Furthermore, how might this apparently complicit popular media format assist in seeding, even instilling alternative environmental attitudes and especially behaviour change?

While not having the answers to such a wishful scenario, nonetheless simply raising such questions and aesthetic possibilities is a good start. Successful environmental narratives that speak to large audiences have to learn to be re-appropriated and actively adopt less serious formats including comedy, parody, pastiche, together with embracing a range of contrasting tones and aesthetic formats to help refine such questions and possible solutions. Even if falling short in this hypothetical communications project and only succeeding in tangentially calling attention to audiences around the gaping paradoxes embodied within such global environmental concerns, this in itself would still be a good start.

Furthermore, mass audiences remain particularly hooked on celebrity culture and seduced by their pleasure-giving affordances. While probably unlikely at least in the short term, imagine if celebrities like James May were to have an environmental 'Road to Damascus' conversion? In any case, while recognising the dangers of wishful thinking, the light green values of such magazine formats are more enticing for mass audiences to engage with, through such easily digestible generic output. This chapter and book strive to call attention to the environmental benefits of actively engaging with a broad range of media genres and stories and their new media affordances – even complicit ones, like this now dated popular magazine series, or recalling earlier greenwashed advertisements like 'Origin Green' discussed earlier in Chapter 3. With some tweaking and a more open-ended educational approach, such formats might help broaden the environmental communications palate and even help reduce the general knowledge deficit, while hopefully signalling the possibility of imagining a low(er) carbon future.

Anecdotally, for instance, I overhear students who remain the target audience for such popular cultural formats like *Top Gear* and its clones speaking of the benefits of an online sharing culture, which includes car sharing. On closer examination, a few argue that May and his *Top Gear* mates are simply sending up their own ideological positions and personifications, being both upper class and middle-aged, and all wealthy English males to boot. Why would they want to do anything to change their current situation, where they can do anything or go anywhere they want? At the same time, it has to be acknowledged that May and many others like him, communicate more effectively with older fans as well as new generations than one might like to admit and therefore need to be understood and appreciated for their nuanced communication techniques and skill-set. Even while knocking an earnest environmental, even frugal agenda embodied by activists like Al Gore, May and other (non-committed) celebrities can help widen the appeal of environmental debates. Imagine, for instance, if such egotistical celebrities could be persuaded to take on-board at least some aspects of the environmental agenda and speak to it. This could in turn greatly assist a broader communications strategy towards reaching a tipping point of environmental transformation and hopefully catalysing broad-ranging change into the future.[11]

Consequently, environmental communication scholars ought to constantly unpick the seductive appeal of such popular cultural artefacts, with their celebrity presenters and decode their use of generic formats to help explore how this strategy might be appropriated, to potentially at least address a counter-discourse that strives to embrace new forms of environmental praxis. Naturally, taking the path of least resistance, most conservative celebrity opinion leaders embrace a techno-optimistic solution to climate change, which does not negatively affect their life-style. Such interventions often mitigate against radical change with regards to adaptation, leaving such adjustments as relatively painless as possible.[12] Yet even some environmental idealists would probably concur that technological fixes have to be part of the final solution. Niche, together with mass media (environmental) narratives have to be constantly co-opted in the communications arsenal, while attempting to get across low carbon messages and solutions, through as many ways as possible, including a more diluted form that is palatable across a wide range of constituencies including non-environmentally committed audiences. All the while of course, remaining constantly aware of the dangers of such complicity and what appears to be mixed messaging and not always highlighting the comparative long-term detrimental consequences. But such risks are worth taking in securing greater audience traction and hopefully reaching a compromised tipping point across various fields of transformational environmental change.

At the same time, environmental communications scholars *must* hold on to their critical role of analysing beyond the present, while providing some hope for the future. There is no scientifically agreed climate change future that will happen in some predefined sequence. Consequently, radical change and transformation will not necessarily occur in a prescribed order. Yet environmental communicators have as a primary goal to get across the urgent need to 'wake up' and prepare to survive into the future, as the current landscape has been moulded for the benefit of our growing dependence on fossil fuels. Modulating and adjusting political and social norms to instigate radical change will require constantly focusing on the primary problem of low carbon energy transition.[13]

Can the media industry create more appealing narratives and audio-visual trajectories that counter some of the negative attributes of such seductive popular texts like *Top Gear* or *Big Ideas*, while noting their ongoing fixation with exciting innovations that promote a 'Business as Usual' trajectory? Or alternatively, by repurposing such appealing tools of audience engagement, can the aesthetic formats and agendas be somehow flipped over, essentially re-adapting their appeal to address more sustainable agendas? The latter tactic is probably the most pragmatic and efficient approach to adopt across a broad range of media output, especially if re-purposed in a more considered and sustainable manner. Using new modes of humour, irony and parody in large measure, as well as appealing to various interactive forms of radical transformation remains an aesthetic possibility, which incidentally new media has been shown to be well capable of embracing and developing. Witness, for example, historical developments around co-opting more positive and critical (re)presentations of gender, class and ethnicity, as explored within contemporary media and film scholarship. Re-purposing such scholarship to address climate change and applying broad ranging textual analysis interventions might in turn suggest ways of transforming this body of critical engagement into explicitly facing up to environmental justice protocols in particular. But even to initiate this process, much work needs to be done towards teasing out and outlining aesthetic and narrative techniques, which can serve this more productive environmental agenda.

Turning from this environmentally suspect magazine televisual format, to a clearly defined pro-environmental documentary, featuring the political celebrity Al Gore and speaking to his much-anticipated sequel to *An Inconvenient Truth*, provides a dramatic contrast across both tone and generic format. Unlike the undoubtedly greenwashed and unreconstructed environmental agenda of May's magazine programme, Gore's serious eco-documentary has a difficult task in striving to highlight how climate change has become an even bigger concern than that presented within his original very popular documentary (see the review in Brereton 2016). But, unfortunately, it appears the sequel broadly labours under the burden of trying to speak to a broader audience and probably not having the requisite aesthetic innovation or even the celebrity performative capabilities that a non-committed celebrity like James May somehow embodies.

Addressing the (American) public's lack of environmental knowledge and literacy: a case study of Al Gore's environmental celebrity profile

In spite of all the untapped new media potential explored in previous sections in this volume, the current evidence of public critical media and environmental literacy is not good. The nature and severity of this educational deficit, one could argue has unfortunately not been effectively communicated through the academy to both educational and governmental agencies. For instance, in *An Inconvenient Truth* which set the benchmark for effective and direct environmental communication, Senator Al Gore alongside many other politicians and scientists seemed to believe that climate change data, if compellingly presented with convincing statistics and some emotional engagement, would change minds on the issue and in turn lead to effective action. Like many conventional environmental scientists, the persuasive celebrity politician still appears to uphold this ‘Scientific Deficit’ model and strategy for effective communication, as further evident in his 2017 sequel.

Overall, *An Inconvenient Sequel* uses less explicit illustrations of science and graphs to get his message across and initially, at least, appears to be provocative enough for mainstream audiences. Nonetheless, a dramatic rising bar-chart showing Chile’s massive growth in the use of Solar Energy solutions to solve their CO_2 emissions remains impressive and has echoes of the original’s famous use of graphs to get its core PowerPoint message across.[14]

In a critical paper titled ‘The Strange Career of *An Inconvenient Truth*’, Finis Dunaway forensically explores how Gore became the world’s most visible carbon warrior and earned cameo roles on popular TV shows and stood onstage with Leonardo DiCaprio to announce that the Oscars had gone green. Gore’s environmental fame also enabled him to become a co-recipient of the 2007 Nobel Peace Prize, awarded jointly to the Intergovernmental Panel on Climate Change, for documenting and popularising knowledge of anthropogenic climate change. As a political environmentalist, Gore is certainly using every measure he can muster to get his message across, and as already mentioned, is certainly on the opposing camp to James May. ‘Gore thus became the only person ever to win a Nobel Peace Prize, based largely upon his role in a movie’ (Dunaway 2015: 259).

‘Be Worried. Be Very Worried’, was the caption from *Time* magazine, in an April 2006 cover story on global warming. The cover features a photograph of a lone polar bear perched on floating ice, gazing uncertainly at the surrounding sea. The polar bear as noted through this study, soon becomes the most recognisable image of climate change, circulating in many media forms (Dunaway 2015: 260).[15]

‘In 39 years, I have never written these words in a movie review’, the legendary film critic Roger Ebert proclaimed, ‘but here they are: you owe it to yourself to see this film. If you do not, and you have grandchildren, you should explain to them why you decided not to’ (cited in Dunaway 2015: 264). Recalling previous environmental icons, environmental scholars tend to agree that *An Inconvenient Truth* frames global environmental citizenship as an emotive project that takes the future into account (ibid.). Like a small number of other environmental trigger points that reference the trope of universal vulnerability, *An Inconvenient Truth* (at least according to Dunaway), nonetheless ignores the systemic practices and spatial inequalities that have shaped – and will continue to shape – the markedly uneven human experience of global warming (ibid.: 269–270).

Gore’s classic won an Oscar and took over 50 million worldwide in box office returns, the sequel unfortunately is another matter. But then political circumstances have radically changed in America. The current Trump-effect has been dramatic, Gore suggests in a (September 2017) interview with Paul O’Callaghan for *ExBerliner* magazine, on the release of his *An Inconvenient Sequel* (directed by Bonnie Cohen and Jon Shenk). Gore recounts that there is a law of physics that sometimes also operates in politics – for every action, there is an equal and opposite reaction. 'And I believe that the excesses and absurdities of Donald Trump have produced a very strong reaction, with progressive movements and climate activism gaining a great deal of momentum'. Now 'we work around him', has become the unofficial (American) climate change policy, according to Gore. O'Callaghan further questioned, if such people who supported Trump are the perfect target of 'alternative facts' and 'fake news', which are so easily spread online. As an early champion of the digital revolution, 'what do you think should be done'? Gore responded to such a leading question by accepting the challenge.

> I think these are growing pains for the internet revolution, and self-correcting mechanisms are now being introduced to empower people to protect themselves against fake news and false information.... For me the fundamental benefit of the internet revolution is that it makes it possible for individuals to use knowledge as a source of power, without having to pay gatekeepers enormous amounts of money. I still think that it offers the hope of reason-based discourse.

Few could question Gore's *bona fides* and credentials as a climate change communicator and tireless global activist. However, reviews have been somewhat critical in their analysis of this documentary sequel. One of the main drawbacks, according to a number of 2017 reviews, is the importance that it places on Gore's work in 2015, during the climate talks in Paris. The sequel all but puts forward the proposition that, as a celebrity superhero, Gore was able to string together the necessary coalition for change, simply through his own determination – recalling the 'Great (white) Man' theory of history. The *New York Times* critic Michelle Nijhuis provides useful comments on Gore's assertion that the 'news has become like a nature documentary, seen through the Book of Revelations', but one wonders is fear enough to spur audiences and citizens into action? As many communication research studies suggest and as also critiqued within this volume, such an enticing strategy is unfortunately not effective of late in helping to radically transform human behaviour, beyond maybe influencing surface and short-term perceptions and attitudes. Contemporary audiences apparently need more sustained and multiple strategies for both short- and long-term transformation, as framed within an augmented and multimodal environmentally mediated landscape.

Furthermore, one must continue to call attention to the fact that media remains only one cog in this evolving process and no matter how climate change is framed, or how sunny or doleful the vision, it's what happens off the screen and off the page in 'real politics' that will decide whether the planet remains habitable. This, of course, is the public stage Gore and other less media savvy politicians remain most at ease with. Media scholars sometimes place too much faith and hope in the representational power of the media, alongside calling on the transformative strength of fictional narratives. While certainly this public forum has its place, the broader political activity together with the business and social agency must always be kept in mind as a fulcrum of transformation, as against the more ethereal transformative power of media representations.

Certainly, environmental scholars have a primary duty to strive to detect effective ways of communicating environmental issues, all the while calling attention to the literature that suggests educating people with facts is not always fruitful, or in not fully recognising the dialogical power of narratives and their emotional engagement. But most worryingly from a communications perspective, scholars remain concerned that celebrities like Gore are simply preaching to the choir (Nisbet 2007). The likelihood of anyone watching the documentary who isn't already sympathetic to Gore's cause seems low, according to several reviews. Such concerns are probably worth teasing out through a broad range of audience studies and responses. Future empirical research will provide concrete evidence to justify a call to extend the environmental template and include a much broader canvas, while accommodating a wider range of media output, as this chapter attests.

From this critical analysis, the directors probably afforded too much time and space to Gore himself as an empathetic personality, activist and world leader and therefore succumbed to the dangers of having their documentary dismissed as a form of celebrity endorsement. Often, on screen, Gore is witnessed preaching his 'new religion', while training his conscripts into how to deal with climate change. Critics have been particularly concerned with his representation as a 'celebrity fixer', recalling the Paris agreement cited above and being portrayed as almost single-handedly persuading the Indian government to sign up to the agreement. Earlier in the documentary, audiences witnessed Narendra Modi, the Indian Prime Minister's curt speech that they wanted a 150-years head-start for the Indian subcontinent, which the West had enjoyed up to now. Effectively alluding to the fact that the industrialised First World had already pumped so much CO_2 into the atmosphere to help establish their economies. The rich West essentially caused the carbon problem in the atmosphere and therefore should fix it themselves. This contention remains a common refrain of Government leaders and even activists from across the developing world in addressing global environmental discussions.

Gore's apparently off-the-cuff response towards recognising centuries of colonial imbalance, included an emotional and at first strange request to look up at the sky. Coincidentally, recording images of pollution across the developing continent was certainly useful and imaginative as an aesthetic ploy to address such well justified criticism by the Indian leader. Gore further highlighted the technological benefits of jumping across a generational technology, as evident by the adaptation of mobile phones in Africa and India, while not having to go through the massive infrastructural investment needed for the earlier telephony revolution. This historical and technological point is dramatised by explicitly calling attention to close-up images of mangled telephone and electricity wires across a mega-Indian city and is designed to reinforce this 'uncluttered' digital transformational message. Powerful rhetorical pronouncements and the use of such iconic images certainly signals new modes of thinking and paves the way for effective transformation – all of which is to be commended.

Unfortunately, however, Gore comes across almost like a salesman for Solar City – an American alternative energy company – persuading its CEO to forego the intellectual property rights for their technology to help the Indian subcontinent and seal the climate change global deal.[16] Thereby, such a strategy appears to avoid the inherently dangerous 'business as usual' route for developing countries, while having to build several highly polluting coal-fired electricity units to help transform their economies, and at the same time pushing the notion of gentle nudging to the extremes of a hard sell. Journalists subsequently looking at the economic record since the Paris climate agreement have asserted that the facts on the ground unfortunately don't substantiate Gore's claim and the move to solar energy in India was very much overstated.

Gore and his team could almost at a stretch be accused by his enemies of creating a subtle form of 'fake news'. Recalling the use of pernicious tactics that many climate deniers – positioned at the far end of the ideological spectrum – are adapting on a daily basis. Fighting fire with fire can be a dangerous strategy. How to protect climate change activists and ensure they, in turn, cannot be accused of misleading the public by exaggerated claims, remains an ongoing struggle. Being totally convinced of the need for environmental transformation, while not having the unlimited resources of the global company and media corporations, or most importantly not always creating audience rapport, remains a major challenge. Furthermore, avoiding the danger of hubris creeping in to the process, while constantly trying to get across their message, environmental activists have a difficult balancing act to negotiate. Most especially facing up to a very suspicious and ideologically compromised mediated landscape and global citizenry with strong even unscrupulous opponents, who have very deep pockets to fight to maintain the status quo, remains an eternal challenge.

The core theme of the documentary sequel – unlike the JMBI's episode discussed above that focused primarily on energy – concentrates on the destructive power of water and flooding, which can be viewed as a direct consequence of climate change. Worries around food, farming and transport, alongside other forms of environmental concerns and especially pollution are kept in the background. Probably, one suspects, this is because the various manifestations of climate change are so numerous, it is often difficult to keep focused discussion and ensure narrative cohesion, if the scope of investigation becomes too broad. Hence, it is most effective in a defined (short) documentary format – as also evident in the previous food chapter and the powerful documentaries that were discussed there – to concentrate on just one aspect, like water or flooding, to help illustrate the bigger picture around climate change. In an early sequence for instance, the audience is privileged in witnessing dramatic images of Greenland and observing breaking icebergs resulting from warmer temperatures. Huge cracks in the ice are seen onscreen, as Gore gets down close to observe, while explaining their global significance. Later, observing from the air, using various modes of transportation from cars to aeroplane, he witnesses the full horror of climate change. Such observation of natural disasters continues by marking the presence of so many flood events, particularly within his homeland America.[17]

But to capture these catastrophic events, stock shots from aeroplanes are used, with Gore observing first-hand out the window, like in a traditional nature documentary.

Unfortunately, there is no even cursory acknowledgement of how air travel has become a substantial contributor to the CO_2 dilemma. All it would require to address this, would be a contextual mention and framed with some humility, that *we* as humans are all in varying ways complicit in causing climate change, especially across the developed industrialised world, with the growth of industrial output and air travel, adding high levels of waste carbon into the atmosphere. Specifically recognising the over-use of CO_2 polluting air travel, environmental documentaries have to become cognisant of such paradoxes, acknowledging how there is probably no other way beyond using drones to capture these spectacular events on film. Otherwise, numerous accusations can be rightly levelled by environmental critics against the creators of such documentaries. Incidentally, James May and his ilk could probably overcome this dilemma relatively easily and effectively with some ironic comment, being less invested than more conventional activists in the environmental project.

Furthermore, the whole documentary focus of *An Inconvenient Sequel* remains heavily American-centric, which is necessary, of course, in providing an antidote to the bellicose climate-denying rhetoric of Trump's new American policy. Gore's approach strives to counterpoint the American President's exasperating decision to pull out of the Paris COP agreement, demanding an additional postscript to the documentary. The strident environmental activist urges (American) viewers to stand strong in spite of shoddy political leadership. Consequently, 'the film stands as a potent reminder of the US president's extraordinary knack of hijacking narratives and serving as a truly inconvenient distraction from the issues that really matter' (O'Callaghan 2017: 30).

After an initial period of growing acceptance, one suspects climate change literacy, driven by a more benevolent proactive Western political will appears to be waning, and the number of Americans in particular (alongside a growing percentage across the world) who accept that climate change is a very serious problem has been diminished. Consequently, and to be fair to the sequel, Gore had to strategically change tack and go 'back to basics', while striving to respond to this serious reversal in environmental attitudes at home, while at the same time constantly highlighting the biggest challenge facing the planet. Some celebrity commentators, like for instance Prince Ea's rapping environmentalism discussed in a later chapter (rather than the more wishful thinking transformation of magazine environmental programmes such as *James May's Big Ideas*), suggest that such political reactionary actions can hopefully serve to actively encourage citizens to finally stand up and become more environmentally active. Most unfortunately, the increasingly partisan and ideological polarising nature of environmental issues has continued to complicate the movement towards climate change acceptance for many Americans – at least as witnessed from outside – while appearing less stark across other regions of the world.

Concluding remarks

Innovative new strategies are always needed to speak to audiences and this is facilitated by calling upon well-established as well as progressive aesthetic formats and alternate voices. The media-sphere needs more and more creative imaginaries to get their environmental message(s) across, beyond the clichés of polar bears on ice-flows and factories emitting toxic fumes, or for that matter succumbing to environmental preaching using provocative scientific graphs and statistics. All the while nevertheless accepting that such historical strategies and well used images continue to have their place in spreading the message to diverse constituencies.

These debates and strategies of engagement however need to be radically re-examined, while re-imagining the possibility of more creative media output and in drawing on audience research, alongside taking into account the development of new images that effectively speak to a range of environmental problems, beyond the over-use of well-worn signifiers and narrative documentary strategies. Learning from pedagogical best practices across other forms of educational literacy, should also include engaging with otherwise suspect and highly compromised magazine formats like *James May's Big Ideas*, from which scholars might uncover innovative ways to address the future of environmental communication.

At the same time, more practical and bespoke audio-visual communication exemplars are needed to be constantly developed. This study believes there is now an urgent need for new creative imaginaries to tell the multiple and often conflictual stories of climate change. Most especially from an environmental perspective, magazine and other mediated formats need to actively address a broad range of generic concerns and inherent difficulties, which are summarised by Tania Lewis and include:

- Putting too much emphasis on individual action – leaving out the need for government or systemic change to address a low carbon future?
- Tendency to blame and shame consumers – while glossing over questions of class and economics and government responsibility.
- Preoccupation with teaching its audiences to adopt implicitly middle-class modes of 'good' consumption and self-surveillance (from actively seeking out organic produce, to purchasing the latest green appliance). Regulating one's consumption and embracing the necessity of inconvenience of green modes of living are offered up as middle-class virtues, to which citizens should all aspire.
- The green mode of address offered up by lifestyle TV, from food TV to the eco-makeover format, including using alternative green energy, is often highly contradictory. This is marked by a desire to offer apparent alternatives to the dominant practices of global modernity and its concomitant embeddedness in a commodity culture, characterised by a growing emphasis on the marketing and branding of lifestyles.

(Lewis 2008: 238)

But as Lewis and others recognise, in spite of all the worries and caveats cited above, such planned yet useful cultural media tend to offer some degree of 'struggle over normative social values around the impact of modernity and consumption' (Lewis 2008: 238) in actively speaking to mass audiences. Also central to such a strategy is a widening palate of story formats, beyond the serious environmental activist model embodied by Al Gore and others. There is an ongoing need to tap into the affective power of popular culture in all its guises. Drawing for instance on a focus around questions of personal responsibility, neighbourliness and an ethics of care, while attempting to wrestle this territory away from conservative populism and begin to reframe personal ethical consumption practices is important. This is especially true 'in terms of collective modes of responsibility and social change' (Lewis 2008: 239). Paradoxically, magazine and new generic formats like *James May's Big Ideas* remains more likely to assist in developing such a potent media environmental taxonomy that will appeal to non-committed citizens and new generational mass audiences, than the more earnest and respected political climate science experts and established documentary formulas.

The media sphere certainly needs to go beyond narrow and purist evocations of the green movement, embedded within crude binary oppositions, inferring that audiences must be either for the environment or against it. *All* citizens have to be encouraged to embrace sustainable levels of environmental behaviour in their own way, or long-term solutions will not be effectively propagated. All forms of media and discursive approaches need to be co-opted in this struggle, including those considered least useful, or not politically radical enough towards developing a constructive environmental agenda. But at the same time this pluralist approach should never be used as an excuse towards dissipating the need for straight speaking environmental documentaries. There is by all accounts no one fixed model of propagating environmental cautionary/productive tales.[18]

Notes

1 As suggested by Lisa Slawter (2008) 'The proliferation of TV channels in post-networked era allows for:

- the potential inclusion of previously excluded perspectives
- the fragmentation and erosion of a mainstream viewing audience
- the fusion of interests such as commercial and activist'

(218)

2 'The Five Psychological Barriers to Climate Change' Per Espen Stokes. Available online at: https://boingboing.net/2015/04/03/the-5-psychological-barriers-t.html.

3 As evident by the green calculator measurement tool used to assess the environmental sustainability of green television and film production – see the project in the UK headed up by the Albert Sustainable TV Production. Available online at: www.bbc.co.uk/mediacentre/latestnews/2017/albert-calculator.

4 In a recent trip to London (April 2018), I visited the British Science Museum where its lack of environmental science appears symptomatic of a current global lack of public engagement with climate change in general. For example, their bookshop had no books on climate change, which was confirmed after checking with the staff. This

appears to provide no pipeline of knowledge and learning to help educate new-generational audiences. While there was a small exhibition on energy on the second floor, this did not call attention to itself. The somewhat underwhelming exhibition nonetheless had useful text provided, including teasing commentary: 'Is this your energy future, exploring Poo power to home energy'. Presumably this was designed to get kids to have a 'yuck' factor to engage with. Typed on the wall was a mention of how can we cut back to clean up our finite energy supplier. Other text mentioned ways around how to use less fossil fuel to power the planet, as this is damaging it greatly because of climate change. A video instillation called 'Donut' by Mike Stubbs, shows a car doing 'donuts' with its screeching wheels, reminding one of the visceral pleasure of fast cars from *Top Gear*:

> 'I need to take risks,
> I love to go fast
> I love to burn rubber'.

5 See, for instance, Sharon Beder's *Global Spin: The Corporate Assault on Environmentalism*. Beder talks of a post-1970s realisation that 'we've been clobbered', recalling how, for example, corporate America formed unprecedented alliances and set up industry groups and disguised front groups to help regain its lost legitimacy, while ensuring that its viewpoint dominated discussions of and helped framed action around environmental issues. By 1990, Beder's research shows that 'US firms are spending about 500 million a year on PR advice about how to green their images and deal with the opposition'. For example, 'General Motors Public Relations helps to make GM so well-accepted by its various publics that it may pursue its corporate mission unencumbered by public-imposed limitations or regulations' (cited in Buell 2004).

6 Clarkson has a long history in baiting what he calls 'eco-mentalists' and pushing a climate change denial mentality, encouraged by his contrarian newspaper column over the years. This includes such comments as 'people will face a rise in their car tax because half a dozen scientists say carbon dioxide emission from cars is causing global warming and that we'll all melt. Its rubbish' (Clarkson 2000) and 'eco-mentalists say we must stop burning oil and gas immediately and go back to living in caves' (Clarkson 2005). 'We are presented with a binary choice: cars or "crazy" environmentalists imposing restrictions on people's lives' (Drake and Smith 2016: 692). Of course, with such unsubtle irony and humour, critics are simply informed they just did not get the joke!

7 Originally a three-part BBC-Open University product which began airing on Sunday 28 September 2008 and whose third episode focused on energy – but now has a new life on the internet and on YouTube. Available online at: https://youtu.be/kTzsv_OwR2I.

8 The awesome power of 'carbon capital' which itself generates rising GHG emissions and is complicit in the overuse of energy has to be radically revised. A major challenge by any measure and unfortunately light nudges designed to incite gentle changes of behaviour will not cut the mustard in itself. But it is a start, as it is envisaged by many that 'capital is expected to solve these issues at the same time' (Urry 2013: 219).

9 Incidentally, they are building another one nearby that can supply all the needs of a city like Seville.

10 Note such dystopic visions are currently being used to critique the spectre of Brexit and its possible effects on the UK economy.

11 See also across later chapters which explore how key influencers and celebrities like Leonardo DiCaprio help shape an environmental discourse.

12 I keep recalling a family member who has had a lifetime devoted to the cut and thrust of marketing and business – which is in some ways very different to the 'rarified environment of academia' – who keeps telling me in very parodic terms when I speak of the need to move to a 'low carbon society': 'good luck with that'!

13 Adjusting the current (Western) model of industrial development, however, which remains predominantly based on how much more material goods and services can be accumulated is a major challenge, which by any measure is pulling in the opposite direction to a more sustainable model of environmental living. Sharing, reducing waste and securing tighter resource management across so many fields, including across media production, has to remain the overall goal and a template for a sustainable way forward.

14 Certainly, there are lots and lots of provocative engaging images in the documentary, but, to my mind, this could be greatly improved and sharpened for audience engagement by maybe cutting out some sections from its 99 minutes overall length. Nevertheless, the prequel and post-rap music video remains effective as a call to arms for new generations.

15 To present the polar bear as poster child of global warming, the media emphasised how the creature's surrounding ecosystem began to succumb to the damaging effects of climate change. 'A polar bear negotiates what was once solid ice', *Time*'s caption explained. 'Bears are drowning as warmer waters widen the distance from floe to floe' (Dunaway 2015: 261).

16 Of course, such an incident would not be out of place within the more complicit strictures and narrative tropes embedded in JMBI.

17 Knowing his core audience, Gore talks of how he was most criticised in the original documentary for the recreation and simulation of a possible flooding of the 9/11 memorial site in New York. But now (unfortunately) he can demonstrate that this really happened through using indisputable visual evidence, alongside showing on-the-ground incidences of severe flooding in Miami, which has become one of the most high-risk sites of flooding, together with Huston in Texas together other global mega-cities currently at high risk across the world.

18 Some efforts are badly needed to uncover how more ostensibly anti-environmental mainstream narratives like *James May's Big Ideas* (as an erstwhile crass example) can be somehow repurposed and re-imagined, while speaking to broader constituencies and encourage not just 'environmental believers' to recognise and learn about a new post-carbon future. Al Gore's more earnest and still powerful profile – in spite of some criticism cited above – suggests that environmental preaching and active engagement with a broad range of environmental issues needs a constant corrective to help speak to a broader constituency of audiences.

References

Brereton, P. *Environmental Ethics and Film*. London: Routledge. 2016.

Buell, L. *From Apocalypse to Way of Life*. London: Routledge. 2004.

Burkholder, K., Devereaux, J., Grady, C., Soliteo, M. and Mooney, S. (eds) 'Longitudinal Study of the Impacts of a Climate Change Curriculum on Undergraduate Student Learning: Initial Results'. *Sustainability*. 2017.

Chayko, M. *Superconnected: The Internet, Digital Media and Techno-Social Life*. London and Thousand Oaks CA: Sage. 2017.

Chen, J. J. (ed.) *Prismatic Ecology: Ecotheory beyond Green*. Minneapolis MN: University of Minnesota Press. 2013.

Clarkson, J. 'They're global dimwits' *Sun*. 22 January 2005.

Clarkson, J. 'Tony's Gas Tax' *Sun*. 25 February 2000.

Curry, A. *Environmental Crisis in Young Adult Fiction: A Poetics of Earth*. Basingstoke UK: Palgrave Macmillan. 2013.

Drake, P. and Smith, A. 'Belligerent Broadcasting, Male Anti-authoritarianism and Anti-environmentalism: The Case of *Top Gear* (BBC 2002–2015)' *Environmental Communication*, 10(6): 689–703. 2016.

Dunaway, F. *Seeing Green: The Use and Abuse of American Environmental Images*. Chicago: University of Chicago Press. 2015.

Gaard, G. 'Children's Environmental Literature: From Ecocriticism to Ecopedagogy' *Neohelicon*, 36(2): 321–334. 2009.

Lewis, T. 'Transforming Citizens? Green Politics and Ethical Consumption on Lifestyle Television' *Continuum: Journal of Media and Cultural Studies*, 22(2): 227–240. 2008.

Miller, T. *Greenwashing Culture*. London: Routledge. 2017.

Nisbet, M. 'Forward'. In Karen Hirsch (ed.) *Documentaries on a Mission: How NonProfits are Making Movies for Public Engagement*. Centre for Social media.org, School of Communication, American University. 2007.

O'Callaghan, P. 'An Interview with Al Gore'. *ExBerliner.com*. 5 September 2017.

Rice, D. A. 'Nonfiction Video Practices as Twenty First Century Liberal Education: The ASPIRE Experiment at UCLA' *Journal of Film and Video*, 69(3): 38–53. 2017.

Slawter, L. D. 'TreeHugger TV: Re-visualising Environmental Activism in the Post-Network Era' *Environmental Communication*, 2(2): 212–228. 2008.

Steger, T. and Drehobi, A. 'The Anti-Fracking Movement in Ireland: Perspectives from the Media and Activists' *Environmental Communication*. 2017.

Stoknes, P. E. *What We Think About When We Try Not to Think About Global Warming: Towards a New Psychology of Climate Change*. White River Junction, VT: Chelsie Green Publishing. 2015.

Thaler, R. and Sunstein, C. R. *Nudge: Improving Decisions about Health, Wealth and Happiness*. London: Penguin Books. 2009.

Urry, J. *Societies beyond Oil: Oil Dregs and Social Futures*. London: Zed Books. 2013.

6 Contemporary Hollywood blockbusters and environmental narratives

Overview

This chapter will analyse a selection of contemporary big-budget and commercially successful films that are read 'against the grain' to help provoke critical environmental engagement. Unfortunately, because of time and space, the chapter cannot provide a detailed contextual and political economic exploration of such randomly selected narratives, which might uncover why such films got made, while nonetheless recounting the makers intentionality – as evidenced for example through the director's commentary as explored in *Smart Cinema* (Brereton 2012). The chapter draws on mainstream contemporary popular Hollywood films, chosen primarily for contrast and include eco-readings of *Passengers* (2016), *Sully* (2016) *vs Deep Water Horizon* (2016) and finally *Blade Runner 2049* (2017).[1] These recent blockbusters can call attention to a range of environmental tropes and interpretations, building on an eco-textual analysis template first exemplified in my 2005 *Hollywood Utopia* study. In particular, such environmental attributes come into play when highlighting emotional and traumatic engagement, sparked directly or indirectly by the growing climate change threat facing our planet. It is argued that these often unconsciously conceived ecological narratives might in turn help audiences call attention to a range of cautionary tales concerning such prescient environmental issues.[2]

As highlighted in films like *The Descendants* (2011), human agency has a major role to play in protecting and preserving our long-term heritage, as we face ever increasingly global environmental struggles. Hopefully, as the range and scale of eco-cinema expands, more creative imaginaries will be created to provoke audiences to engage with such complex issues that call upon emotional, ethical, legal, scientific, biological and other environmental literacy protocols and agendas, all of which coalesce around 'doing the right thing' for the environment. Coping with the cycle of life and death remains emotionally challenging, especially recalling evidence of increasing levels of climatic disaster and much more difficulties being forecasted into the future. Environmental scholars speak of how all citizens should be encouraged to think and feel more holistically, creatively and long term for the very future of our species and the planet. Because of their broad appeal, mainstream Hollywood big budget narratives remain a powerful formula, which in turn help project such prescient learning processes and dramatise long-standing mythical stories and tropes that speak in varying ways to mass audiences across the globe.

Narrative structures which can help support environmental literacy and citizenship

Antonio Lopez argues that an ecocentric model of media ecosystems helps audiences to rethink how they actively engage with media. Rather than act as mere 'audiences', 'participants', 'users' or 'prosumers', we are instead constituted as 'members' that are part of 'communities' of a much larger media ecosystem 'that encompasses global communications networks across the biosphere' (Lopez 2015: 152).

Unlike economic citizenship, which is primarily characterised by passive consumerism and market fundamentalist logic, cultural citizenship can alternatively be greened, among other ways by an eco-ethical orientation. Such a focus entails recognising that 'human beings live in a more-than-human-world, of which they are only one part'. Whereas economic citizenship is considered de facto anthropocentric, or human-centric, a green cultural citizen alternatively is ecocentric by embodying sustainable behaviour and cultural practices that in turn help shape and promote ecological values. This corresponds with Mitchell Thomashow's holistic concept of ecological citizenship (Lopez 2015: 153). But how can filmic representations inspire audiences to appreciate much less embrace such deep ecological values? This question of course remains an ongoing preoccupation embedded within eco-film scholarship.

Key environmental attributes which need to be detected and developed through a process of environmental literacy investigation are set out by Frances Lappe's model of the 'Eco-Mind'. This includes highlighting human capacity for cooperation, empathy, fairness, efficacy, meaning, imagination, creativity and plasticity. Such ecological intelligence is akin to systems thinking, because it is based on seeing patterns and relationships from small to large scales and understanding interrelations between them. For instance, Donella Meadows asserts that the purpose of systems theory is to perceive the relationship between structure and behaviour (see Lopez 2015: 157).

Eric Beinhocker (2006: 126–127) in turn reckons that stories can assist with appreciating relations between macro and micro concerns, as embodied by overall structures and individual behaviour and have become an evolutionary mechanism for inductive reasoning. As the ancient Greek philosopher Plato once said, 'those who tell stories rule the society'. In turn, stories are vital to society because the primary way we process information is through induction. Induction is essentially reasoning by pattern recognition (Booker 2004: 141). 'Humans particularly excel at two aspects of inductive pattern recognition. The first is relating new experiences to old patterns, through metaphor and analogy-making.' While the second suggests that humans 'are not just good pattern recognisers, but also very good pattern completers. Our minds are experts at filling in the gaps of missing information' (Beinhocker 2006: 127). Fictional narratives work very well in helping to fill in such gaps and audiences can discover and certainly learn a lot around environmental knowledge, by simply engaging with the process of storytelling. This process appears especially true with the pervasive appeal of apocalyptic stories, embedded within the science fiction genre.

Apocalyptic end of the world narratives and science fiction

The apocalypse might well be embedded in our collective consciousness, since it functions as myth, teleology and as a narrative device across popular culture. Indeed, the apocalyptic imagination remains particularly protean, for it performs many functions and assumes many guises in the contemporary world. These are exposed through forms of 'endism' and apocalyptic narratives.[3] All of which are much beloved especially across many dystopic science fictional films.

As a clearly coded cautionary format, a number of science fiction films almost instinctively reflect the urgent need to respond to upcoming catastrophic changes to our planet, which can in turn be ascribed to long term concerns around climate change in particular. Such films serve as cautionary tales around what might happen if humans don't begin to seriously address major environmental and related ethical issues. For Ulrich Beck, who coined the phrase risk society, environmental hazards can never be eliminated through the use of technological knowledge, although it can be anticipated. Beck claims that we now live in a highly troubled risk society that continues to find answers to these particular environmental problems, using the logic of the nineteenth century heuristics. Mark Smith for instance cogently summarises Beck's ideas:

> while the hazards of a technologically driven society in the late twentieth century penetrate every region and level of society, human beings remain wedded to the responses to environmental degradation, which were more appropriate in the nineteenth century.... The scale and scope of human impacts upon the environment produce a range of complex and unanticipated consequences, which cannot be contained effectively within the earlier guarantees and safety mechanisms.
>
> (Smith 1998: 94)

Speaking to the dominance of climate change as *the* global challenge and witnessing major weather events across the planet, environmental risk ethics have become even more evident, pervasive and engaging across a wide range of contemporary sci-fi films including, for example, *Oblivion* (2013), *Interstellar* (2014), *Arrival* (2016) and *Passengers.* Sociologists such as Adam, Beck and Van Loon have identified how contemporary environmental problems are not always easily accessible to the senses, giving them an 'air of unreality' until 'they materialise as symptoms' (Allan *et al.* 2000: 3). Hence the urgent need for the creative imaginary of powerful mass mediated narratives, especially film, need to be called into action to help bring such destructive processes to light.

Adam (1998) has argued further that the very nature of environmental problems – including issues around sustainability, adding a further layer of complexity – happens over a long period of time and are often invisible and thereby not appreciated or understood by general audiences.

Nonetheless, Hollywood has made environmental apocalypse perversely attractive. Recalling some environmentalists who, as part of their arsenal of communication strategies, have adapted the Old Testament concept of doomsday with its perverse appeal, 'waking us from our humdrum existence'. In spite of some of the psychological research examined in earlier chapter, audiences may still be emotionally attracted to apocalypse, according to Steven O'Leary, through a desire for consummation, narrative closure or absolute knowledge (O'Leary 1994: 66). So, the ever-present set-piece of apocalyptic weather and global destruction, effectively exhibits the totalising and sublime power of wild nature that is violent, chaotic, somehow amoral and beyond human control. In many revenge-of-nature films, wild nature effectively functions by transgressing ethical norms, as audiences witness nature 'getting its own back' as it were, for its maltreatment at the hands of human beings. Engaged (environmental) media spectators apparently take vicarious pleasure in the destructive forces of nature, from the safe distance of their cinema seats, for example, eco-films like the much written about *The Day After Tomorrow* seek out particular forms of audience identification, not only with regards to wild nature, but also paradoxically with the forces of civilisation that try to control that nature.

Holmes Rolston 111, in an ethical examination of human duties towards alien constellations, alongside protecting our natural flora and fauna back on earth, suggests that *we* must first appreciate all as individual containers of creation, since humans live in an 'inventive universe'. We 'confront a *projective nature*, one restlessly full of projects, including stars, comets, planets, moons, radically changing sky-scapes and also rocks, crystals, rivers, canyons, seas' (1986: 155). The life in which these astronomical and geological processes culminate in, remains impressive, if only from a vicarious 'cinema of attractions' perspective, where the historical apparatus and the science fictional imaginary has over time discovered new ways of representing and speaking for the natural world. Nonetheless, this process is clearly out of kilter between humans and nature, as evident by the planet's ongoing difficulties. Facing up to global apocalypse, while taking into account its effects on nature, as well as ultimately on humans, the environment remains the most pervasive series of interconnections and forms of creative imaginary that such films actively seek to engage with and warn against.[4]

Such a dramatic and growing back-catalogue of 'end of the world' films expresses such elemental historical and social anxieties, coalescing around the interpersonal and human focus of their narratives, while highlighting how everything is connected (Morton 2018). Yet acknowledging the climate scientist Mike Hulme's worries that such discourses of catastrophe are always in danger of tipping society into a negative, depressive and reactionary trajectory, one can still counter that such narratives might at the very least, call attention to such climatic catastrophe and concurrently suggest fruitful forms of reflection, if not necessarily always serving as models for future action.

While such broad-based textual analysis of science fiction in particular is useful in teasing out a range of environmental/ethical aesthetics and concerns, environmental scholars must continuously question what impact these films might have in creating an explicit awareness of climate change. Scholars like Allison Anderson for instance point out that certain issues gain public legitimacy 'through their capacity to become icons or symbols for a wide range of concerns that people can easily identify with' (1997: 5–6). Others, like Julie Doyle, go on to warn that in the case of climate change predictions, often the lack of explicit visible evidence of such problems, makes it difficult for the issue to be linked to an established set of symbolic imagery (2007: 133). Commercial exploitation of environmental issues across mainstream cinema nonetheless helps to make such issues easily identifiable, and, probably if only on a surface level, makes visible these complex climate change problems. There is always a further danger of environmental issues becoming simply topical and populist, while finding it difficult to sustain their longevity, as the broad ecological church pulls in various directions. Nevertheless, such commercially driven audio-visual media can project useful scaffolding and models for often disaffected audiences to latch onto, by highlighting provocative modes of dramatic engagement with global environmental concerns, as evident through an analysis of *Passengers*.

Passengers: sublime evocations of nature in space

Marshall McLuhan once famously said 'There are no passengers on spaceship earth. We're all crew'. The problem is that, whether one comes from privilege or lives a life of struggle, within today's media and consumption driven commons we are all unfortunately invited to act like passengers and must accept the invitation. Yet large segments of the world's population are encouraged to only have market-driven relationships with the environment that rarely challenge people to think or act like crew, much less in being co-responsible for their own ecosystem and habitat as passengers.

This ecological ambivalence can be explained in no small measure by the fact that 'today's media institutions have not been leveraged in any meaningful way to break free from a cycle of passive recognition and weak ecological, consumer-based action' (Murphy 2017: 145). At the centre of such 'business as usual' structural practices, which unfortunately does not encourage any meaningful form of critical media citizenship, is a discourse of perpetual growth, expressed through the media market, which Patrick Murphy and others refer to as the 'Promethean discourse'. Scarcity, we shall recall, is understood by the Promethean's as an economic, not an ecological challenge (Garrard 2004), with new resources sought out, found and taped into, only when they are needed. Such ever-present global tensions are evident for example in the back-story of the most recent high-profile eco-parable *Avatar* (2009) (see the reading in Brereton 2016).

Many scholars suggest that fantasies of apocalypse are both a product and a producer of the Anthropocene. Although images and narratives of contemporary environmental apocalypse have usually been understood as politically regressive and affording evidence of post-political distractions. Hopefully, science fictional tales like *Passengers*, help to demonstrate that a more environmental reading of such tensions is both possible and useful. Apocalypse tells us that the human, as currently configured in the Anthropocene, have become an ideal universal subject, who remains energised through fossil fuels and who has been elevated to a position of ecological mastery – but this cannot continue indefinitely.

Starring Jennifer Lawrence as Aurora Lane (of *Hunger Games* fame) and Chris Pratt as Jim Preston, while directed by Moten Tyldum, *Passengers* represents a conventional science fiction love story that nonetheless raises several ethical and philosophical concerns around 'doing the right thing' when the possible survival of the human species is at stake. The story encapsulates the trajectory of many recent big budget sci-fi movies that explicitly call attention to the global common good. The detailed exposition of a spaceship in flight with all its many corridors and open spaces, coupled with the dramatic manoeuvres automatically carried out on the outside of the ship, are probably the most engaging and sublime aspects of this futuristic film-world. In some ways such scenarios remind one of a synthetic, game space environment.[5]

At the outset, the story focuses on a very sophisticated space-ship called Avalon carrying 5,000 passengers and 259 crew-members on its 1,200-year voyage to a distant colony/planet known as 'Homestead Colony', in an effort to escape the apparent environmental ravages back on earth. Notions of a new frontier are raised, recalling the pervasive drive of the dominant historical western and science fictional genres and discovering a new benevolent habitat to begin constructing communities and societies again. This pervasive and common scenario calls to mind the ongoing dystopic fears of massive climate change and the resultant environmental catastrophe on the planet.

The main protagonist, Jim Preston, is a mechanic who likes to work with his hands and probably this is the primary reason why he is chosen to wake up 90 years before officially programmed to do so, when a catastrophic failure is registered on the ship's instrumentations. Reminiscent of the blue-collar Tom Cruise character in *War of the Worlds* (2005), who drives a crane and thereby has enabling skills that prove useful when extra-terrestrial chaos comes to planet Earth, Preston's skills are similarly needed in this futuristic storyline to help all the passengers escape from mortal danger.

Preston spends over a year on his own trying to find his rhythm, while getting to know his space-ship environment. Unfortunately, there remain restricted spaces, including the crew and technical areas that cannot be breached – recalling ever-present class divisions and gated communities back on earth. Hence finding solutions for the survival of the space ship requires another agent to help gain access to these areas and avoid impending disaster. A senior officer who is also magically brought back to life, Captain Gus Manauso (Lawrence Fishbourne) serves an assistant role that is common across mythical sagas and science fictional narratives – most famously witnessed in the *Star Wars* franchise.

Furthermore, like a Robinson Crusoe generic figure and recalling the Tom Hanks character in *Castaway* (2000) where the hero relieves total boredom and isolation by conversing with a painted volleyball, Preston here uses the assistance of a quirky cyborg bartender called Arthur (Michael Sheen) to serve as friend and companion. Nonetheless, conforming with hetro-normative drives, he secretly desires a female partner to make such a radical transformation palatable and his life meaningful. So, after much trepidation, he decides to 'wake up' a beautiful passenger called Aurora, who Preston admires from afar back on earth. After a slow courtship, they naturally fall in love, before the bar tender lets the truth out. Aurora cannot accept such deception, going so far as to accuse him of murdering her. Divorce proceedings ensue, recording a full gamut of melodramatic fighting and banging around the lifeless spaceship, scenes that would not be out of place in a conventional soap opera storyline.

But the primary directive of survival of the whole ship and its crew suddenly becomes the number one concern of the film. Triggered by several incidents involving technical malfunction and dramatised to great effect during Aurora's efforts to gain physical release through swimming – strangely recalling Timothy Morton's notion of the hyper-object (also exemplified in other chapters) as an objective correlative to help bring the laws of physics into dramatic relief. She is suddenly visualised losing gravity, with all the water in the pool being reconstituted into a single bubble. Meanwhile, attempting to alleviate the emotional tension and make up for his selfishness, Preston constructs a 'natural tree' for his beloved in the foyer of the ship. This instructive form of response and capability for true organic creativity provokes a defining environmental creative imaginary within the narrative – certainly beyond the human convention of giving flowers as a token of forgiveness. This action I suggest further calls to mind the intrinsic potency of (first wave, romantic) environmental art, which can intuitively speak to audiences and call attention to human desire to tap into the therapeutic value of nature. A dramatic process which intrudes on an otherwise synesthetic spaceship environment.

Furthermore, this *mise-en-scene* and spectacular visualisation of constructed raw nature, marks out and embodies an active manifestation of the celebration of the cycle of nature, which can eventually draw the couple together. His drive to actively relate to the potency of nature and illustrate his deep biophilic connection with the therapeutic power of living plants, help to push the narrative towards simulating a powerful environmental allegory. This creation of nature on a large scale certainly provokes an environmental appeal towards active engagement with nature conservation and is reminiscent of the smaller scale real flower/shrub, highlighted within the narrative trajectory of *Wall-E* (2008), which also becomes the modus operandi of environmental transformation and learning (see eco-review in Brereton 2016). Organic and wholesome nature is worth engaging with and fighting for, such narratives imply, as they help to bring meaning to life. This certainly appeals to the therapeutic power of nature, broadly encapsulating the overall eco-message of such big-budget storylines.

Dramatic tension is ramped up further, when their technologically sophisticated habitat becomes even more precarious, on discovering that their ship will blow up if they don't fix a major breach to the surface of the spaceship. To solve these technical difficulties, Preston must venture out into open space to attempt to open a locked outer-door and literally dissipate the problem. Like many Hollywood heroes, he has to selflessly and instinctively do what is best for the greater good. The protection of the spaceship/planet is all that finally matters. All the while, focusing on and contemplating the 5,000 + other inhabitants who will lose their lives if he does not act immediately. This protection of the common good scenario is somewhat reminiscent of the celebrated 'hill of beans' speech embedded in *Casablanca* (1942). Reflecting a critical juncture and narrative touch-point across so many classic Hollywood generic narratives, when it really matters, heroes have to step up to the mark and act totally selflessly for the greater good of humanity, while ignoring concerns for their own safety. Such altruistic generic tropes and ethical messages have to be re-told for different generations, as witnessed later in an analysis of *Sully* and further call attention to much needed global problem-solving and resolution strategies across so many environmental cautionary tales.

Finally, once the immediate dilemma has been resolved and the spaceship is back on its steady course, a final moral conundrum has to be addressed. Preston discovers that the medical pod can double as a sleeping dock, but it can cater for only one person at a time. Such magical technology therefore leaves them with a difficult decision – who will be chosen to survive their final journey and destination?

Again, an ingenious humanitarian and egalitarian solution is enacted (spoiler alert) – fast forward to 90 years later, when the here-to-fore unseen crew and passengers are finally waking up across the ship to the biophilic natural sensation of a well nurtured garden. This celebration and recognition of (organic, even Old Testament, First Wave environmental agency) natural growth remains a potent visual signifier that would not have been possible without the heroic action of its two crew members, who literally devoted their lives for the common good. The newly awakened crew members, together with the filmic audience, discover a verdant rich and fruitful natural environment spread all across the decking area, as the accompanying music swells in memory of the lovers' sublime sacrifice.

Using an environmental lens to interpret such a scene, while focusing on such a revelation, life, love, art and a concurrent fruitful environment – almost a new allegory of the Garden of Eden – have left their mark for new space-generations to draw on. This is a clear and simple illustration of the development of a cogent eco-metaphor and a clear application of the potential of the creative imaginary to highlight the sublime appeal and therapeutic power of raw nature and of reconnecting the intrinsic strength and necessity of human sacrifice to fight global destruction. By all accounts such imagery evokes the romantic and selfless love of nature which our two lovebirds held for each other, in spite of their earlier disagreements.

One surmises, of course, that this creation of a 'natural garden' and beautiful holistic environment was more than enough for the two heroes, realising they had taken on a death sentence (no more than the rest of all human nature), while, at the same time, left to their own devices to achieve full self-actualisation. Love, and presumably the therapeutic powers of gardening, is all they wanted and needed in life to make their precious time together fulfilling and rewarding. They chose each other over long-term (artificial) survival on their own. This selfless act, the embodiment of a romantic drive across centuries of artistic representation, is poetically visualised and actualised by the organic natural landscape and habitat they created with their own hands and left behind, is enough in itself. This environmental edict celebrating the intrinsic benefits of leaving a healthier habitat for those that come after us (while maybe not registered at the same dramatic level of heroic fulfilment, as marked out and witnessed in *Casablanca*), serves to highlight the benefits of adopting a proactive environmental ethic in the face of global destruction, and at the same time stepping up to the mark to take firm dramatic action to assist a space-ship (earth) from succumbing to an unstoppable environmental catastrophe.

But, one wonders if audiences recognise or even appreciate this implicit environmental message, or if there is enough explicit signifiers of such an ecological sensibility and interpretation to support this eco-reading? By all accounts there is much provocative philosophising within the script and the diegesis of the narrative helps to support this interpretation, while encouraging viewers to ethically engage with the overall message and hopefully decide not simply to be just a 'passenger in life'. Rather the tenor of the narrative trajectory privileges how it is more important to manage one's finite time on earth and to make good choices when faced with environmental-ethical dilemmas.

The film calls to mind much of the environmental learning and ethical protocols of stewardship needed to support sustainable living, while dealing with a whole range of crises, all of which can affect spaceship earth, if we don't take proper care of its future welfare. The storyline provides a relatively straightforward and uncritical unpacking of such concerns, as one expects from mainstream media, which nevertheless allegorically speaks to the primary need to both nurture and protect our habitat. This ongoing struggle has to be highlighted as a major challenge for publics to engage with. However, only primary audience research can help uncover if such environmental messages are explicitly, or at least subconsciously, decoded in this way, or if this or other science fictional blockbuster films might actually connect across new registers of environmental engagement. But at least in calling attention to and highlighting such a pro-environmental framing, this textual analysis approach can assist audiences in re-reading and evaluating such pleasurable dramatic narratives, while helping to discover the growing potentiality of deep environmental connectivity. All of which is very different to a contrasting, real life heroic disaster story we now turn to, framed from a bird's eye view.

Sully vs *Deep Water Horizon*: a birds-eye view of climate change (a cautionary tale of flying from an ecological and sustainable perspective)

As noted in *Sully*, after 9/11 it was a long time before *we* had a real life 'good news' story about a plane crashing in New York, or for America for that matter. In this docu-drama, audiences witness Sully's (Tom Hanks) endless nightmares about his plane crashing into buildings, replicating the horrors of 9/11. This nightmare speaks from the conscience of an experienced pilot of over 40 years, including earlier serving in the American Airforce, where he tested new planes. This expositional memory is sparked out jogging in the cold night air, while waiting for his judgement on the accident and passing a ferry-port with military planes on board. As he quips to his co-pilot, he is being judged on these 208 seconds,[6] as a measure of his whole flying career. The 'everyman' character Hanks plays is somewhat reminiscent of his earlier role in *Captain Phillips* (2013), where, rather than transporting people, his character was helping to transport a huge volume of containers on a super-tanker to keep the means of global capitalism in full flow.

Meanwhile, the pilot's wife is also worried, but more so regarding financial concerns it would appear, recalling the threat of losing his livelihood and not being able to fly any more. Consequently, the couple envisage losing their holiday home and not having enough income for their old age. Surprisingly, having his own flight training/safety company – as part of the neo-liberal solution to support income top-up – is not considered that unusual as a financial safety net. Sully certainly has a lot of duties and responsibilities to juggle with, while the well-sought-after career of commercial flying apparently has its own financial insecurities and worries. What does this apparent economic reality suggest about living and working in contemporary America?

The story, of course, hangs on whether the pilot made the correct judgement with his emergency landing on the Hudson river in the cold weather of 15 January 2009, rather than attempting to make it back to a nearby airport. The simulator apparently suggests the pilot could have made it to a safer landing, but later it is revealed that this preferred solution was only achieved after 17 attempts, using the simulation technology.

Recalling an eco-reading of an extreme weather film *Twister* (1996) (see Brereton 2005) and the tension raised between intuitive experts, who rely on professional and most especially intuitive and lived experience, as against those who only have the latest technology to rely on to make instant life/death decisions. While in charge of so many lives, Sully has to make a catastrophic and instant decision, which will dramatically affect the lives of all his 155 passengers. Heroically, of course, Sully made the right choice, as audiences know in advance from the real-life news story. This is further affirmed by the wreckage of the plane, when the engines are discovered to have been totally destroyed by a bird attack. Again, this does not tally with what the apparently scientific and empirical simulations were suggesting, which in turn informed the crash investigators' risk-analysis and lead to a questioning of the performance of the pilot.

Like so many heroic classic Hollywood storylines, the investigators affirm that the positive solution to the crisis was simply down to the 'x-factor' of Sully, as expert and highly experienced pilot, who instinctively deducted the right thing to do in such a split-second crisis situation. Drawing on a long tradition, the whole film seeks to celebrate this mode of professional and heroic action. But of course, our everyman appears more humble, affirming that he would not have been successful in this near disaster without the help and assistance of his co-pilot and all the crew and passengers, not forgetting the boats which helped quickly and efficiently take the passengers off the river. By any measure, this constitutes a success story, which ideologically serves to support the American Dream of safe and efficient travel, where communal solidarity can also help save the day, underpinned by the leadership qualities and fast reaction times of trained specialists.

Alternatively, one could argue from an environmental perspective there is another story being told, while using a very different lens (Schuurman and Nyman 2014), by reading the narrative against the grain, as a deeply buried allegory concerning society's excessive preoccupation with easy and available air travel and which coincidentally remains a major cause of climate change. Recalling, for instance, the failure of high-technology, coupled with faulty human reaction from a previous century, as portrayed most effectively in *Titanic* (1997) (see Brereton 2005). Using a similar eco-perspective and lens, the birds replace the unseen iceberg, which further serves to question, if not critique, what in climate change terms might be considered the smoking gun, namely the high-tech carbon-wasting modes of commercial air travel. The proliferation of air travel remains one of the most concentrated CO_2 pollutants on the planet, yet this reality is rarely if ever critiqued within mainstream cinema.

Counter-intuitively maybe, one can further note that there were unfortunate casualties, namely the unseen Canadian geese destroyed in the incident. Adam Smith's (2016) journalistic report of the accident notes how:

> at 3.27 pm the plane was travelling at an altitude of 3,000 feet directly over the Bronx and at a speed of 250 mph, when it hit the birds travelling at around 50 mph and being sucked into the planes' two CFM 56–5B engines which flame out and fail immediately, though miraculously do not disintegrate. The animals are instantly liquidised, transformed into a thin mist of what is known to air crash investigators as 'bird slurry'.

One wonders if our culture is so anthropocentrically pre-conditioned, with a fixed set of human values, that the slaughter of flying birds by a passing aeroplane does not even register across any ethical level of measurement. The bird strike is simply articulated as an unintended consequence and a freak accident of nature. Consequently, the bird strike is dismissed as a dangerous nuisance that has to be simply managed and controlled.[7]

Unlike, for instance, Hitchcock's *The Birds* (1963), which could be considered an early environmental allegorical story, the non-individualised creatures of this real-life tale are ostensibly portrayed as pests and not overtly individuated or represented, much less elevated to become malevolent agents in a psychological disaster movie, who might intentionally strive to attack humans (or their flying machines). Instead, putting a positive spin on the dilemma, they are simply 'God's creatures', but in the wrong place at the wrong time.[8] Yet pushing the nature/human hypothesis further, who has the innate right to traverse so much air-space besides the birds, since this is *their* natural habitat. Nonetheless, across contemporary ideological discourse, these majestic creatures in this aerial habitat (unlike in a nature documentary most notably) are simply regarded as pests to be destroyed, as they negatively affect the well-ordered modes of human flight transportation.[9]

With so much sophisticated and sensitive radar available, one wonders why such sensitive technology cannot accurately record the birds and their natural flight and avoid entering their flight path. Probably, one supposes, because the birds' path is unpredictable and somewhat erratic and thereby impossible to detect in real time when collisions are about to occur.[10] Most pointedly, however, for the purposes of this eco-reading, bird flight involves no (criminal) carbon discharge into the atmosphere, unlike, of course, human air travel. Nonetheless, few, if any, environmentalists are openly calling for a total ban on aeroplane travel to help radically reduce our carbon footprint, much less to secure the safety of the bird population. At the same time, the plight of the birds within such scenarios serve up a cautionary allegory, which is seldom called attention to when dealing with existing and long-established high carbon technology emitters like air transport. Unlike most notably the fear of bird strike which is prescribed as a defined environmental high risk and precautionary principle laid out within Governmental regulations and specifically taken into account when refusing planning applications around the construction of wind farms, for instance, which are designed to reduce the human carbon footprint. Paradoxically, it would appear, new clean alternative energy sources of low carbon emission and solutions to climate change, like wind or even solar farms are considered high risk to the environment, because of the dangers of bird-kill. Meanwhile existing modes of wasteful CO_2 intensive technology like air travel, is not even on the radar.

Deep Water Horizon

At the same time, one can further contrast this reading with an explicitly environmental storyline, which highlights one of the few evocations of nature and birds being harmed in a fictional mainstream film. *Deep Water Horizon* (2016), directed by and starring Mark Wahlberg, is a true story about a BP drilling rig off the American coastline, which became the biggest oil spill disaster in American history.[11] In one particular micro-scene, audiences graphically witness the tangible effects of the disaster, when a very scared pelican, totally covered in oil, unceremoniously lands on a nearby ship's engine room and causes havoc, before dying. Yet this disturbing image of nature-in-distress simply remains an expositional back-story incident, to frame a fully fleshed-out family drama of survival of its blue-collar oil-rig workers.

Wahlberg's screen daughter is so proud of her oil-rigging father who is pumping oil – with her knowledge of the cult children's programme *Dora the Explorer* – that audiences are given a basic history lesson of how 300 million years ago dinosaurs were squished together and the pressure got greater and greater, resulting in the creation of precious oil deposits. The process of extraction is further simulated – for ease of audience engagement and understanding – using a carbonated tin of Coke and a rod being pushed into the can, to illustrate an explosion. By all accounts a simple visual signifier to explain the underpinning premise of the (dirty) oil exploration industry. But, of course, in the real-life scenario such deep natural pressure and technical measurements coupled with human sophistication goes horribly wrong. This is allegedly posited because of BP's greed and not taking enough precautionary measures before extracting this powerful energy source, a scenario which was aggravated by being well behind schedule in the work completed. The actual process of extracting oil and the resulting CO_2 emission difficulties for the planet is not critiqued, however, unlike the way, for instance, fracking is examined in Mat Damon's more effective portrayal and cautionary tale *Promised Land* (2012).

Like in *Sully*, birds and nature generally take a back-seat to the consequences of an oil rig exploding, which is dramatically felt by all the unfortunate workers. Such environmental cautionary tales are simply pushing the need for more regulation with its requisite checks and balances around risk management. However, this storyline is not necessarily designed to question the very *raison d'etre* of oil extraction and its detrimental effects on our planet. The final reading in this chapter goes back to examine the ever-popular science fiction genre, recalling the longstanding academic and cult popularity of *Blade Runner*, which has recently been rebooted and has taken on a fresh ecological angle.

Blade Runner 2049 franchise: replicant farmers – a new mode of environmental learning and literacy

With a duration of two hours and 40 minutes, the long-awaited sequel certainly takes itself very seriously. Recalling the original masterpiece which has served to underpin and illustrate so much postmodern theorising and philosophising (see review in Brereton 2005), this reboot pushes the franchise in new environmental directions.[12] Most importantly for the purposes of this chapter, the sequel has much to say about the state of environmental decay facing humanity and might even posit new forms of active engagement, which are needed within such contemporary and future world-views.

The iconic *Blade Runner* (1982) probably needed a reboot for a new generation, recording a new sophisticated aesthetic and other special effect resources delivering high production values in this re-working of the franchise. Denis Villeneuve, as director, certainly felt the heavy weight of the original on his back and worried if his new version might measure up against its legacy.[13]

Environmental engagement and learning, I suggest, are explicitly afforded through the authenticity exuded by the first Replicant, Sapper Morton (Dave Bautista) to be 'wasted'. He is a proud and generally silent farmer, working peacefully in futuristic aqua-farming while harvesting (GMO) worms in his plastic covered water-tanks. Not the usual nostalgic image of nurturing farming that comes to mind. But, for a post-human society, this is the nearest one can probably get to organic and self-sustaining food production. A big bulk of a man – who is a very different farmer to the psychotically driven Bull McCabe from the Irish film *The Field* (1990) – he is, if left alone, happy fulfilling his stewardship duties, with no interference and keeping his deep memories intact. While in his humble hut, his meagre meal is being slow-cooked. Agent K (Ryan Gosling) waits for him to finish his duties before they begin a deadly tussle, breaking much of the sparse furniture and even some of the flimsy dividing walls of the building. Almost in passing, he unconsciously mentions how he is so happy, having had a 'miracle' in his life. On further investigation, this turns out to be the closest manifestation of a holistic biological community – namely a (human) child to love and cherish. Such natural human desires should have remained scientifically beyond the hope of synthetic machine humanoids or replicants.

On closer observation of the iconic (dead) tree, positioned outside the farmer's homestead, a coffin-like box is discovered underneath the soil.[14] When surveying this strange (organic) tree from above, deep surface analysis reveals the bones of a woman. One presumes she was buried there as part of some human ritual and ceremony. Through greater magnification of the bone evidence, using high-tech machinery back at base, it is revealed that the corpse had undergone a crude C-section on birthing a child. By any assessment this was indeed a miracle, as replicants can't physically reproduce like humans. Only someone 'born' of woman (as the Shakespearian riddle goes), rather than 'made', can have a soul or produce human children, which remains the oft-repeated mantra. Later, we discover that Deckard and the replicant Rachel from the original story has produced a daughter Ana. K's boss pronounces how this capacity to procreate is indeed a revolution, a game-changer. Consequently, such knowledge of the apparent evolution of a synthetic replicant species has to be buried from public consciousness and hopefully wiped out, if the ideological system which reifies synthetic rather than organic natural organisms is to maintain its long-term balance and security.

Creating synthetic memories from a nurturing landscape

Like the farmer Sapper Morton – who is conventionally recognised within eco-narratives as caretaker and steward of the earth – audiences are finally introduced to the real-life daughter Ana, who is protected by her family. Using her extraordinary powers, she has creating an artificial digital world where she can safely live out her desire to actively engage with nature, but without any subsequent danger. As Zizek caustically states in his analysis, however:

> Living in her isolated world, unable to survive in the open space filled with real plants and animal life, kept in utter sterility (a white dress in an empty room with white walls), her contact with life is limited to the virtual universe generated by digital machines, she is ideally positioned as a creator of dreams; she works as an independent contractor, programming false memories to be implanted into replicants.
>
> (Zizek 2017: 4)

One could conclude that Ana remains the antithesis of the 'down and dirty' farmer discussed earlier and is certainly not a part of the raw natural world, which unfortunately she can never experience first-hand, but only realised as a gameplay simulation.

Reminiscent of *Logan's Run* (1976) or the more evocatively real construction of an idyllic form of nature in *Passengers* explored earlier – recalling of course its burgeoning green habitat constructed for future generations – Ana simply 'imagines any world' she desires, having enormous powers of mimetic synthetic creativity. She spends her time like a natural scientist/botanist, engaging first-hand with her reconstruction of nature. But as in so many science fiction simulations, it's to no avail, as her imaginative exercises remain a simulacrum. Nonetheless, such creative digital activity keeps her happy and pure. Usually only organic nature and authentic environmental systems can achieve this long term and sustained engagement. So, when she finally meets agent K and is asked to assess the provenance of his timber horse to confirm if it is a 'real memory' or not, she cries out while focusing on its natural authenticity. As in *Wall-E* (2008) or *Waterworld* (1995), recalling so many other similar science fictional narratives including the original *Blade Runner*, access to, much less ownership of, some pure piece of organic/non-synthetic matter or life force is prized above all else, within such a digitised and synthetic high-tech world.

All of which works towards the continuing drama, both as a nostalgic loss of an authentic nature and at the same time stimulating a desire to protect what humans have over the centuries created. Coincidentally this tension remains a key tenet of environmental media education, involving respect for an authentic and organically-formed planet. Ana cannot, of course, be harmed or affected by her creations, remaining locked inside a hermetically sealed world she cannot escape. Meanwhile, the rest of the inhabitants have to suffer the endless 'hard rain' and chaos of an environmentally decayed habitat, presumably initiated by man-made disasters and climate change. This is reminiscent of the constant precipitation shown in the original film, coupled with the noir-like smog and highlighting of garish signs blotting out further dereliction.

In one such isolated place, Gosling (Agent K) witnesses the full horror of this re-imagined (post-colonial) technologically-driven world, which has, over time, totally corrupted the harmony and balance of the natural world. Young children are being used – like in contemporary Third World sweat-shops – to break up the used computer circuit boards and other forms of electrical waste from their over-technological world that in turn cannot bear the weight of such decay. This powerful sequence effectively calls attention to the growing environmental problem of e-waste (see Miller 2018; Cubitt 2017), coupled with the ongoing global human injustice of Third World manipulation of its undervalued citizens. This almost in itself justifies the texts' badge as an ecological tale and memorable parable around environmental mis-management. The owner of the sweat shop incidentally does not see himself as a modern-day slave-owner, but rather as being a 'good capitalist', involved in the re-purposing of otherwise waste material, while constantly creating new wealth for society.[15]

But after much struggle and angst, as Agent K tries to discover the meaning of it all, the iconic protagonist gets shot. Reminiscent of Karl Reed's classic *Odd Man Out* (1947), or James Joyce's famous short story and John Huston's adaptation of *The Dead* (1987), he ends up having Propp-like served his narrative function by delivering the 'original Deckard' (Harrison Ford) to his long-lost daughter. Apparently dying in the snow on the steps of the building, Agent K. appears to sacrifice himself – as in a Christ-like gesture to create the final closure – again reminiscent of so many fictional heroes across the Hollywood *oeuvre*.[16]

Concluding remarks

Can the potentiality of revolutionary fervour be extended to include environmental security – as evident for example in passivist narratives like the World War Two true story *Hacksaw Ridge* (2016), or the transcendental trauma of deciding how to be an effective environmental activist in *First Reformed* (2018) – recalling a science fictional fantasy like *Passengers* discussed earlier. The more everyday heroism and documentary realism of *Sully*, with a captain protecting his passengers by making a safe landing, remains somewhat ecologically suspect when taking into account the unwitting victims of the story, namely the Canadian geese. Or, for that matter, the oil-covered pelican in *Deep Water Horizon* fleetingly represented in the real-life saga of a huge oil disaster, which highlights the ecological dangers of environmental pollution when big business ignores regulations and strives to take short-cuts to ensure financial success. Finally, in *Blade Runner 2049*, the lynch-pin character of K has fulfilled the ultimate altruistic goal of communal living, by suffering for others, so that a better life can be gained for all.

As in so many Hollywood narratives discussed in this and other volumes, Agent K, and to a lesser extent the romantic heroes witnessed in *Passengers* and *Sully*, can serve as exemplary heroes for mass audiences to embrace and vicariously empathise with as they engage with and address a range of sometimes elusive environmental issues. In spite of several caveats, such celebration and glorification of heroes at least can be read as a useful starting point for developing more robust environmental communication and imitative forms of pro-environmental behavioural performance and active engagement.

Notes

1 While being somewhat disappointed at the lack of explicit environmental fictional narratives being produced of late – which probably speaks to how low on the popularity scale such issues are held within the Hollywood gate-keeping community. Scholarship is badly needed to explore the political economy of the American industry, especially in teasing out the contextual back-story of media output, while accounting for the narrow range of films being produced. Thankfully, this situation is less consistent across other national film-making industries, which alternatively support a broader range of environmental storylines, but unfortunately such tales are less well distributed and thereby less popular across global cinema screens.

2 In ecocinema studies this is what MacDonald calls the process of promoting 'cinematic experience that models patience and mindfulness' (2013: 19).

3 Available online at: www.google.com/search?biw=1440&bih=736&tbm=isch&sa=1&ei=P45PW-niO8nCgAab_q-QCQ&q=%28Hart+and+Molba%2C+2009+Ecocatastraphism&oq=%28Hart+and+Molba%2C+2009+Ecocatastraphism&gs_l=img.3...43174.48997.0.49758.17.17.0.0.0.0.79.951.17.17.0....0...1c.1.64.img..0.0.0....0.CjR8E2VFou4.

4

> Everything is made out of dirt and water, stellar stuff, and funded with stellar energy. One cannot be impressed with life in isolation from its originating matrix. Nature is a fountain of life, and the whole fountain – not just the life that issues from it – is of value.
>
> (Holmes Rolston 1986: 197)

5 Such sophisticated and extensive use of outer space paraphernalia most certainly draws comparison with Kubrick's seminal *2001 A Space Odyssey* (1968).

6 See report by Adam Smith 'Sully: Miracle on the Hudson' in the *Telegraph*, 22 November 2016.

7 See, for instance, episode 7 of *Ozark* on Netflix which dramatises the somewhat strange, loner boy Jonah, who becomes an expert in 'drug laundering' following in his father's footsteps and earlier refused to sign a school petition on same, because he argued the illicit drug industry was 'essential' for the American economy. Furthermore, he wanted to be allowed to shoot starlings as a pest – having heard a documentary on his computer which affirms that such 'pests' were imported into America initially by someone who wanted all the birds mentioned in Shakespeare to have a new home in the US. This produced disastrous results for native and indigenous birds – many of which almost reached extinction as a result. The worst apparently was starlings, who attacked crops and killed other birds, as well as being a major nuisance for aeroplane engines apparently.

8 See Mary Douglas' famous definition of 'pollution' as matter or 'nature out of place' (1966/2000)

9 Consequently, many airports have 'anti-bird' units who send up gunshots and use various forms of remedial action to stop birds from flying into the path of planes which actively suck them into their engines as they use so much power and energy – but again I ask – who has the right to be in this space more than the birds!

10 See for example the romantic self-actualisation allegories like Richard Bach's novel and subsequent film *Jonathan Livingstone Seagull* (1973) – or a host of nature documentaries which celebrate the majesty of such unimpeded flight.

11 Yet nowhere as near as bad a pollutant for the oceans as is the scourge of plastics – as highlighted in environmental series like *Blue Planet 2* and addressed in provocative websites like www.stumbleupon.com/su/3mQH28/:18RvuU3__:Ph4!XjNY or www.alansfactoryoutlet.com/how-long-does-it-take-plastics-to-break-down.

12 *Sight and Sound* (November 2017) interview with director Denis Villeneuve – he was 14 when the original was made. He has also directed *Prisoners* (2013), *Sirario* (2015), *Arrival* (2016) and felt it was important to have Ridley Scott on board with this

reboot, who in turn co-wrote the script. In an interview with James Mottrom, the director spoke of how he wanted to add deserts and other sites beyond an urban L.A. while keeping the spirit of 1980s with Atari video games still present. When asked if the film is still pursuing themes of the world on the verge of environmental collapse? The director affirmed that it is in the background, the sub-plot. Like the original, it's in a world where the environment and eco-system has collapsed, where there was a major change to the climate and the earth is slowly dying (27). But that is all, nothing else. It certainly I would conclude does not posit any 'useful' solutions.

13 The tension and continuing fixation with human ontological values is further explored in the sequel. At first sight, the new version feels almost like a prototypical right-wing moral fable, similar in ways to *Hacksaw Ridge*, including at a stretch a 'Pro-life' homage to the ultimate sanctity of life at all cost, which incidentally many deep environmentalists would embrace. This reading will instead focus on the sequel's evocation of new modes of (GMO, genetically modified) farming and the ability to re-imagine nature and creative habitats as a short-hand but potent ecological signifier.

14 Incidentally, the representation of a tree remains a fantastic eco-image, recalling the way various aspects of the natural world, including how flowers and trees are represented in *Passengers* discussed earlier and also feeds directly on our romantic/environmental fixation with the 'creative imaginary'.

15 Incidentally, the resilient global franchises are constantly present and presumably still need to be protected, including the ever favourite 'Coke is good'.

16 Recalling the leader of the replicants who want 'justice' and the prospect of revolution to fight for their rights, pronounces 'the only true liberation, is finding a global struggle that is good and worth dying for'.

References

Allan, S., Adam, B. and Carter, C. (eds) *Environmental Risks and the Media*. London: Routledge. 2000.

Anderson, A. *Media, Culture and the Environment*. London: UCL Press. 1997.

Barbara, A. *Timescapes of Modernity: The Environmental and Invisible Hazards*. London: Routledge. 1998.

Beinhocker, E. *The Origin of Wealth: Evolution, Complexity and the Radical Remaking of Economics*. New York: Random House. 2006. Available online at: www.neweconomicthinking.org/downloads/HEEDnet%20Seminars_Eric_Beinhocker.pdf.

Booker, C. *The Seven Basic Plots: Why We Tell Stories*. London: Continuum. 2004.

Brereton, P. *Environmental Ethics and Film*. London: Routledge. 2016.

Brereton, P. *Hollywood Utopia: Ecology in Contemporary American Cinema*. Bristol UK: Intellect Press. 2005.

Brereton, P. *Smart Cinema: DVD Add-ons and New Audience Pleasures*. Basingstoke UK: Palgrave. 2012.

Cubitt, S. *Finite Media: Environmental Implications of Digital Technologies*. Durham and London: Duke University Press. 2017.

Douglas, M. *Purity and Danger: An Analysis of Concepts of Pollution and Taboo*. London: Routledge. 1966/2000.

Doyle, J. 'Picturing the Climat(c)tic: Greenpeace and the Representational Politics of Climate Change Communication' *Science as Culture*, 16(2): 129–150. 2007.

Garrard, G. *Ecocriticism*. New York: Routledge. 2004.

Holmes Rolston 111 'The Preservation of Natural Value Systems'. In Eugene C. Hargrove (ed.) *Beyond Spaceship Earth: Environmental Ethics and the Solar System*, pp. 14–182. San Francisco, CA: Sierra Club Books. 1986.

Lopez, A. 'Putting the Eco into Media Eco-systems: Bridging Media Practices with Green Cultural Citizenship'. In R. Maxwell, J. Roundalen, and N. L. Vestberg (eds) *Media and the Ecological Crisis*, pp. 152–176. London: Routledge. 2015.

MacDonald, S. 'The Ecocinema Experience'. In Stephen Rust, Salma Monani, and Sean Cubitt (eds) *Ecocinema Theory and Practice*, pp. 17–41. New York: Routledge. 2013.

Miller, T. *Greenwashing Culture*. London: Routledge. 2018.

Morton, T. *Being Ecological.* Cambridge MA: MIT Press. 2018.

Mottrom, J. 'Crimes of the Future' *Sight and Sound*, 27(11): 25–27. 2017.

Murphy, Patrick *The Media Commons: Globalisation and Environmental Discourses*. Chicago: University of Illinois Press. 2017.

O'Leary, S. *Arguing the Apocalypse: A Theory of Millennial Rhetoric*. Oxford: Oxford University Press. 1994.

Smith, A. *Film/Sully*: 'The Miracle on the Hudson: How it Happened'. *Telegraph*. 22 November 2016.

Smith, M. *Ecologism; Towards Ecological Citizenship*. Milton Keynes: Open University Press. 1998.

Schuurman, N. and Nyman, J. 'Eco-National Discourse and the Case of the Finnhorse' *Sociologia Ruralis*, 54(3): 285–302. 2014.

Zizek, S. '*Blade Runner 2049*: A View of Post-human Capitalism'. *The Philosophical Salon*. 2017. Available online at: https://thephilosophicalsalon.com/blade-runner-2049-a-view-of-post-human-capitalism/.

7 An environmental analysis of post 9/11 American televisual series

Overview

Using the work of John Parham and others to explore how new modes of environmental literacy can be applied to popular television series that have dominated world screens, this chapter will develop an ecological reading of an episode from *The Sopranos* and *Homeland*, both of which have been very successful. While at the outset such televisual output and storylines appear to have little to do with green issues, nonetheless one can illustrate how they do relate to global environmental and ethical concerns, making such complex issues both engaging and even educational for mass audiences. These readings will specifically focus on themes of waste disposal and food consumption, coupled with personal revenge, alongside dealing with global injustice and international terrorism. Complex environmental themes raised in the series will be decoded using 'against the grain' or oppositional modes of textual analysis – a strategy which became popular in film and media studies from the 1970s onwards, recalling feminist readings of patriarchal and conformist Hollywood films (Modleski 1991).[1]

Rorschach's psychological inkblot test is more usually used to uncover an underlying mental disorder, by requesting patients to make associations around what comes to mind when looking at a number of inkblot drawings. But in this instance by replacing drawing with an audio-visual stimulus, the test could potentially be re-purposed to help illustrate mass audience appeal for archetypical and psychologically flawed characters within these series. Coincidentally at least this test helps to frame an eco-reading process that viewers might apply towards unpicking such affective and multi-layered representations – recalling Barthes' exploration of the polysemic text. Audio visual media can present a similar, albeit more sophisticated exercise to the inkblot test, which can in turn help towards uncovering latent environmental, alongside psychological preoccupations embedded within the diegesis of these character-driven storylines. While the concrete presence of an environmental theme, embedded within such popular narratives, remains open to contestation, nonetheless this psychological interpretative approach can assist in highlighting, even greening, media studies.

While textual analysis has been consistently carried out on fictional film as a site of ecological engagement, television is often considered the medium most resistant to green ideas (see McComas and Shanahan 1999). Nevertheless, according to Parham (2016), considering the extent to which digital or web-based television has become so popular of late, it might offer new spaces for environmental exploration. This chapter strives to spell out how the new immersive spaces embedded within a number of televisual series, both through their unique affordances of extended screen time and high-quality application of aesthetic space, have incorporated complex environmental themes, which arguably over time might become embedded into viewers' everyday experiences and consciousness. An examination of these high-quality popular television series (which can incidentally be viewed not just across conventional television viewing platforms, but also online and through 'binge viewing' on box-sets) facilitates and often encourages the development of elaborate and challenging themes and interventions, having both the extended time and big production budgets necessary to support such innovative programmes that coincidentally also speak to broadly conceived environmental issues. Audiences, alongside environmental communication scholars, can actively engage with such polysemic texts and their rich narratives in ways that allow them to tap into deep emotional, as well as cognitive environmental triggers, which help to raise awareness of ecologically salient issues, especially for citizens not necessarily committed to such engagement, as these two case studies might suggest.

From waste disposal to surveillance: *The Sopranos* and *Homeland*

Written by David Chase, *The Sopranos* provided a compelling insight into the psyche of a new breed of violent Mafia types, running on American HBO cable television from 1999 to 2007 and which became hugely successful with international critical reviews and massive audiences. The series served to dramatise society's inability to resolve personal, much less more global conflicts, be they political, environmental, social or personal issues. Dana Polan (2009) most notably calls attention to the series' fixation with waste disposal, with the main protagonist affirming from the start that their public front has to be protected, 'garbage is our bread and butter'. The 'equation of waste and sustenance well captures how the show interweaves accumulation and the production of by-products of refuse and rot' (Polan 2009: 137). Only a small number of storylines explicitly highlight such issues directly, however, recalling most notably a cautionary episode that dramatises disposal of environmental waste, by secretly dumping it into a pristine lake, thereby highlighting the need for environmental conservation and protection.

Meanwhile, another globally successful series *Homeland*, which began on the American Fox channel in 2011 and continued into 2018 (the seventh series being broadcast in 2018) offers a psychological profile of the heads of major governmental CIA operatives and their Muslim, terrorist and Russian adversaries, dramatising some of the most controversial, yet highly realistic, global terrorist and security scenarios portrayed on television.

Coincidentally, the renowned scholar Richard Rorty posits the view that in recent decades America is not 'in search of a public philosophy' but, rather, is 'in search of a moral identity' (in Sandel 1984). As a result of the increasing fragmentation and atomisation of (Western) society and what some consider as the disintegration of the human spirit, as well as the apparent breakdown of coherent political meta-narratives, there seems to be a growing demand for popular mass media to fill a philosophical, spiritual and ethical void, one which could also be seen to fulfil a broadly ethical environmental role that promotes long-term sustainable responsibility (Brereton 2016). This growing desire remains particularly noticeable, following the polarising of political and philosophical debate after the September 11 atrocities, alongside more recent attacks on all sides, as global conflict escalate, which in ways appear to mirror more long-term environmental tensions. Both of these series in their different ways speak to such global tensions.

Homeland has certainly become a global success story and a defining American political thriller series.[2] It stars Claire Danes as Carrie Mathison, a Central Intelligence Agency officer with a severe form of bipolar disorder. The series focuses on the growing power of surveillance and the need for security, which includes various forms of environmental security concerns, some of which are triggered by the kinds of global inequalities and injustices that are frequently missing from anti-terrorist discourses. Carrie's unpredictable mental and psychic imbalance in particular, serves as a motivational justification for some of her more irrational behaviour and responses. To help tease out the multi-modal storylines and by extension the latent ecological residue of such representation as also evident in *The Sopranos* – while accepting the difficulties of decoding and reading such mega-series through one predetermined lens – it is revealing nonetheless to zone in on one insightful if random episode from both of these series to tease out some specific environmental issues and concerns. But first of all, it is helpful to outline some of the unique affordances of such long-play televisual series as part of a growing phenomenon within new media production and audience reception.

New televisual affordances and the evolving capability of digital platforms

A quick survey of the literature indicates that none of the televisual scholarship focused on these popular tales have even alluded to possible engagement with pressing environmental concerns, suggesting an apparent resistance to green ideas (Parham 2016). This is not surprising however, as there is still only a handful of media and film scholars who actively seek out environmental readings and interpretations of mainstream film output, much less strive to highlight such perspectives in their textual analysis across various other media formats (see Rust *et al.* 2013; Hiltner 2015; Weik von Mossner 2017).

Environmental televisual readings bear with them the potential to tease out the extent to which quality digital or web-based media can offer new spaces for exploring a range of environmental themes. Extended televisual narratives certainly harbour 'a capacity to unravel the complexities intrinsic to ecological engagement' (Dunleavy 2009: 222). Dunleavy further suggests that this new phenomenon (she calls 'series-serial'), which promotes 'narrative experimentation and interest in visual spectacle', essentially can be put down to both 'the creative control expected of the "star" writers and producers that have created it, and to the additional licence for experiment that their niche orientation; high-end budgets, and "must see" objectives have warranted' (222).

While the spatial and visual properties of these new televisual outputs open up clearly defined aesthetic possibilities, extension in screen time remains a primary way television audiences can develop a close familiarity with a complex theme like environmentalism. Both *The Sopranos* with its focus on contemporary gangsters and ostensibly 'doing the right thing' with regards to waste disposal of all types, alongside *Homeland* and its fixation with surveillance and technological fixes used to address (environmental) security issues, demand extensive (and often intensive) time commitment by its global and multi-faceted audience. Frequently the chief protagonists avoid 'doing the right thing' for themselves, their family, much less their community. In this way, audiences are encouraged through engaging with and experiencing such long-play ethical narratives, to oscillate between identifying with and alternatively critiquing the main characters' often nefarious action and behaviour patterns. Much can be uncovered by exploring how such tantalising and hypnotic dramatic engagement is precisely actualised, through their respective gangster and global security genres, while at the same time being re-calibrated to evoke a range of environmental concerns.

Affirming this approach, Stuart Hall has long argued that the message of television drama 'intersect with the deep semantic structures of a culture' and take on additional, more active ideological dimensions (1973: 12). One could suggest, this is particularly true for touchstone and globally populist television drama like *Homeland* and *The Sopranos*, which not only reflect the growing importance of environmental, cultural, social and political anxieties, but also actively and creatively reflect back those anxieties, sensitising and shaping them for society at large. To begin to appreciate these latent powers and tease out how audiences might co-opt new forms of environmental literacies to help frame and promote such green-tinged interpretations, the opaque nature of the Rorschach test is called on as a short-hand decoding metaphor to help illustrate the power of such multi-layered modes of psychic characterisation and thematic engagement.

Pushing audience engagement to the limit: recalling Rorschach's psychological test

The metaphor of Rorschach's psychological inkblot test can, at its most basic, serve as a simple reference point to help explain the process of engaging with such crime drama, 'because it highlights the extent to which anyone can make what they will out of such a figure' (in Short 2005: 189). Such televisual shows can therefore be allegorically read as an extended audio-visual Rorschach test, helping to promote psychological and even, I would argue, a form of green learning. Whether audiences are on the left or the right of the political spectrum, or alternatively marked out as actively engaged with environmental issues or not, all viewers can find much to latch onto and engage with through such multi-layered narrative and memorable *mise en scene(s)*. Environmental concerns can be seen to be at least allegorically evident in *The Sopranos* and *Homeland*, through the way the characters frequently speak across ethical as well as broadly green issues and become 'larger than life' and emotionally affective cyphers in their seductive appeal to mass audiences through their often-violent characterisations. Meanwhile, at a more direct register, such representations of psychotically framed characters can also function as ethical conduits, helping audiences tease out various political and cultural tensions and post 9/11 worldviews that further resonates across various global environmental concerns.

At a surface level, *The Sopranos* appears to be the very antithesis of what might be regarded as an earnest ecological model or exemplar. Recall its highlighting of the main protagonist Tony Soprano's (featuring the late James Gandolfini) over-consumption of food, his flash use and consumption of material goods and his general arrogant predisposition, all of which serve to isolate him from the long-term consequences and strictures of environmental concern. Tony and his family are thus positioned as far away from a frugality-driven, environmentally-aware, and even literate section of society, as one is likely to come across in the real world. However, reading against the grain, such overt selfishness and excess directly calls attention to itself, all the while signalling such conduct as being the antithesis of good and virtuous ethical behaviour.

Meanwhile, *Homeland*, through an investigation of various forms of surveillance, speaks most directly to a growing international and political fear of the Other, while also exploring how gendered and psychologically compromised human agents are used to respond to the growing evidence of terrorist threats, as well as dealing with environmental injustice, with varying dramatic outcomes and final resolutions. This trans-national, multi-layered narrative has in some ways become one of the defining audio-visual maps for our globally insecure time and could be re-interpreted as signalling the need to also further address human security from a broadly ecological perspective. This is particularly evident through the psychologically complex layering of Carrie's personal life, which (like Tony Soprano) can be read as highly ambiguous, like a Rorschach test image that serves to gauge and measure audience responses, as her behaviour becomes more and more erratic (see Pirkis *et al.* 2006). Her psychological disintegration is (re)presented through earnest looks at the camera and catching reflections in mirrors, which in turn invite viewers to join-the-dots and get inside her head, recalling a bespoke psychological test. All the while viewers are trying to decode her interior tensions and sometimes erratic behaviour.

Teasing out the interior struggle of the main characters and discovering what drives them within a Machiavellian world-view, audiences uncover the broader cultural and political agenda being pursued by such nefarious, yet always engaging three-dimensional protagonists. Again because of the extensive diegetic time/space being used up and portrayed on our television screens – involving literally hundreds of hours of dramatic viewing – the main protagonist's cognitive, psychological, affective and ethical mindset cannot be easily discounted, much less minimised as an instrument and conduit of disposable dramatic fiction. The protagonist's psychic residue and excess (drawing on narrative developments from Screen and Feminist film theory from the 1970s) is further rounded out by the cult/fan audience, committed both by time and creative effort to faithfully follow the ever-evolving trajectory across ever-expanding story-lines. This close audience engagement with such psychologically challenging protagonists, recalling feminist and psychological-film engagement with the symbolic register of screen characterisation,[3] can metaphorically at least be extended to include the workings of a psychology test. This, in turn, can be re-applied to the exigencies of uncovering environmental triggers and strategies that are needed in addressing human behaviour change to help tackle global environmental issues.

Audiences experience a (disloyal) thrill in seeing Tony Soprano, alongside more benevolent protagonists like Carrie, or the staider Saul (Mandy Patinkin) in *Homeland*, who selfishly do whatever they want as they flout conventional rules. Such larger-than-life figures appear infinitely better equipped for survival and problem-solving than mere mortals like *us*, not only in terms of their effortless ability to assimilate, but also through the sheer physicality of their presence (Short 2005: 198). Nevertheless, as Vivian Sobchack[4] has noted in another context regarding film and media engagement, it takes a little pain to bring us as audience back to our senses. Such provocative, rich quality televisual series evolve and weave a web of often competing tensions through character identification and exploring wide ranging storylines, while the viewer is continually oscillating between fascination and repulsion for these powerful, yet often compromised characters, who constantly display complex and psychologically believable personas.

What is particularly interesting is the way such psychologising of an institutionally violent family can also allegorically serve to reflect a deep-seated discontent, even neurosis, within the American psyche, which is both aesthetically pleasurable and at the same time ethically bankrupt. Alternatively, if only by default, *The Sopranos* draws heavily on more positive communitarian and other socialising markers, especially around protection and family blood ties, all the while privileging the ever-present moral imperative of 'doing the right thing', as audiences are tempered by the suspect world-view of these pathologically-driven main protagonists. The psychologising of such behaviour almost functions like an 'empty signifier', or alternatively recalling the Rorschach inkblot test, by encouraging viewers to decode narratives and images as they see fit and take pleasure from what satisfies their individual and idiosyncratic psychic needs.[5]

Unlike more conventional conflicts and the ongoing global War-on-Terror discourses, dealing with climate change – the predominant red line environmental issue – appears to address much more nebulous challenges and is gauged at a greater scale of complexity, with the enemy less easily detected. Furthermore, it is extremely difficult to build consensus around the radical changes necessary to take on board, much less take on the fight against environmental catastrophe in particular, with no clearly defined (scapegoated) enemy available to blame. At one remove however, the life of a fictional mafia family provides a fruitful morality fable and back-story to help trace these and other changing national and global tensions and even throw some light on contemporary events, while demonstrating how 'everything is connected' (Morton 2018), which remains a well-worn truism embedded within deep ecology. At the same time, the pervasive neo-conservative manifestation of American power politics, alluded to by Rorty earlier, is frequently contested by *The Sopranos* and most especially within *Homeland*, through what can be described as a tentative endorsement of an inclusive form of communitarianism. At a stretch, this can be further explored as an idealised political model which can help promote deep environmental modes of sustainability. This ideal is set up through a dramatic series of conflicts and structural oppositions, where the political or utopian dream involves finding harmony across various structural levels of society.

The most contentious structural opposition embedded within both series is set up between erratic and psychotic type behaviour, as against what is categorised as normal activity. As Joshua David Bellin observes in a chapter on the 'monstrous minds' in fantasy film and mental illness: 'though a white person cannot become black, or a man become a woman (at least not inadvertently), a sane person can, without volition, acquiescence, or even awareness, become insane' (2005: 39). Like the relatively easy ontological transformation of human to cyborg in cinema, actors can make the transition to 'madness' relatively easy, by simply modulating their performance to take on an extra-diegetic level of engagement. Audiences, therefore, have to constantly and actively read such behavioural pathology to evaluate dramatic human responses. Tony Soprano easily carries out this transformation with his 'mild' manifestation of depression, a condition which underscores the whole series. But this form of psychological transformation, coupled with a fatal flaw register of vulnerability, is taken to another level with the persona and performance of Carrie in *Homeland*. As Susan Sontag explains in *Illness as Metaphor* (1978), we are all fascinated by physical and mental illness. 'Illness is the night-side of life, a more onerous citizenship. Everyone who is born holds dual citizenship, in the kingdom of the well and in the kingdom of the sick' (ibid.: 40). Such dramatically flawed main protagonists and characters further speak to a contemporary society that no longer holds on to predefined meta-narratives and apparently no longer has a coherent set of social and culturally shared values to which one can adhere.

Nonetheless, all forms of psychological analysis, together with deep environmental engagement and active agency crave the desire for a renewed meta-narrative, which one could suggest places environmental concerns around justice and sustainability at the central nexus of contemporary political and cultural debates. Audiences also crave popular cultural mediations around global problem solving, while alternatively questioning the rules of 'normal selfish behaviour'. They can thereby begin to envision, if not connect with, radical transformation, which is required to kick-start and hopefully integrate a deep ecological transformation. In any case, as in all great drama, such dramatic protagonists constantly perform on this tightrope between normality and dysfunctionality, which in turn helps to show the way for audiences to respond. To illustrate some of these issues, a discussion of two specific episodes from each series will now be examined, recalling their unique use of immersive narrative and psychological engagement.

Vengeance and retribution: over-consumption of food and the American dollar

In the narrative climax of this first episode in series four of *The Sopranos*, 'For all the Debts Public and Private' (directed by Allen Coulter and written by David Chase), Tony reveals information which is used to consolidate his relationship with his nephew Christopher (Michael Imperioli). This involves bringing him to where a retiring police party is taking place in order to finally expose his father's killer. Tony explains how the policeman was 'useful' until he retired but was now expendable. Such sharing of secret knowledge serves to deepen their familial blood ties and viewers learn that Tony secretly wants Christopher to become his successor. Later, following an ongoing piece of theatrical business within the series, the retired man is being brutally interrogated, and Christopher's father avenged, while in another diegetic space Tony is happily consuming his lunch.

Food plays a major part of the pleasures of the mafia protagonist in this series (see Lavery 2002; Edgerton 2013). But rather than measured healthy eating of organic food, which would support an environmental and sustainable sensibility, Tony is frequently observed in typical domestic scenes, attempting to chill-out after a 'hard day's work' with a cream Sunday or a late dinner from the fridge, while indulging in front of the TV, watching a soppy sentimental period movie. None of these consumption choices are good for his weight, or his general physical or even psychic health, much less the environment. In the fictional old movie world-views in which Tony indulges, however, the moral compass is more easily codified and defined, thanks to clear narrative closure and black and white ethical clarity. Alternatively, handling complex environmental problems that need sophisticated multi-layered (even green coloured) responses don't work so easily in dramatic terms, since protagonists (and scriptwriters) prefer clear-cut solutions to conflict, which in turn produce a different story. Consequently, environmental problems like dealing with waste, human or otherwise, are often unceremoniously disposed of. The environmental justice edict encapsulated by the 'polluter pays', never mind taking into account the long-term consequences for the planet of such forms of spoliation, remains far from the mind of most gangsters, if not also from more mainstream everyday practice of business enterprise.

As Christopher is savagely beating a confession out of the newly retired policeman, the now bloodied and frightened victim confesses that he is being set up. 'Who cares', the mafia underling responds, enjoying the cathartic blood sport of unadulterated vengeance, thus appearing to negate any ethical treatise on justice – environmental or otherwise – while actively affirming the raw desire for revenge serving to obfuscate any other impulse. This trope has become a continuing narrative preoccupation within Hollywood and also one might infer across American foreign policy, as is clearly evident in *Homeland.* Afterwards, Christopher rifles the dead policeman's wallet and extracts a single dollar bill, as a token and a trophy. Such a shockingly memorable incident is reminiscent of the ever-present threat of irrational violence carried out for its own sake and constantly experienced in the extensive Hollywood back-history, recalling the 'savage scalping' of Native American Indians, which was critically addressed in revisionist Westerns like *The Searchers* (1956). Or as further witnessed in contemporary (anti)war movies like *Apocalypse Now* (1979), which present both hunter and hunted as morally equivalent (see Brereton 2005).

Later, apparently in a predetermined act, Christopher ends up placing the iconic paper bill on his mother's fridge door, alongside other memorabilia of her past life with his father. The iconic centre of food consumption in the home, the humble fridge, becomes metaphorically tainted as an active site of memorial and revenge, serving as a reminder of how their lives have been reduced to retribution and violence. His mother's life, we discover earlier has been soured after her husband was killed and she became depressed, apparently exacerbated by the trauma of 9/11. Not surprisingly, Christopher does not reveal anything of these events to her. That his family has been avenged is all that apparently matters.

The camera finally, self-consciously and memorably, zooms into an extreme close-up of the face and eyes on the dollar bill in a closing image of the episode, metaphorically and ironically signalling 'an eye for an eye', recalling, I suggest, an ambiguous Rorschach test image demanding notice, while allegorically at least alluding to how cold-blooded 'justice' has been meted out. All of this exposition remains the very antithesis of community and holistic justice values that actively need to be upheld if deeper forms of environmental values are to be privileged and become normalised across society.

Yet it appears that mass audiences are vicariously happy to observe from the outside what daily preoccupies such corrupt criminal families, with for instance studies finding middle class business types particularly enjoying such nefarious storylines (see Lavery 2002). Many commentators on the series continue to speculate on how audiences appreciate and relate to, if not identify with, such a family and their extremely violent actions, alongside effectively engaging with their unashamedly amorality within contemporary society, much less caring about what might be coded within environmental, communitarian or other registers. But as, for instance, recognised across all great Shakespearian drama, by highlighting the extremes of immorality, injustice and most especially in dramatising psychological imbalance, audiences are alternatively encouraged to re-imagine and re-connect with what is deemed positive and truthful towards supporting healthy ethical behaviour. This includes recognising constructive aspects of environmental ethical practices and 'doing the right thing' for audiences, across several aspects of society and even striving towards being 'at one with nature'. Certainly, a big interpretative ask. Nonetheless, this form of polarising dramatic representation and oscillation through active audience engagement with a broad range of psychological tensions, at the same time recall various global environmental issues, remaining a utopian hope and latent promise of great audio-visual art.

Meanwhile, *Homeland* moves away somewhat from the tight focus on family and gangsters and alternatively addresses an extensive array of global tensions and dissonance, while dramatising violent governmental conflict. By highlighting global terrorism and related political tensions, the series successfully addresses complex inter-connections within and across political and security systems, including environmental conflicts over oil and water played out across the world. In brief, using a critical eco-textual reading, the series co-incidentally speaks to global power politics and related climate justice issues, highlighting, for instance, debates around the use and abuse of natural resources, which becomes a major trigger point for action that is most evident across various destabilised regions in the world. Most pointedly, various forms of surveillance are called upon to observe, decode and interpret these conflicts. This dominant trajectory ultimately serves to privilege and promote a high-tech solution for political as well as climate change problems.

Homeland: the psychology and ethics of surveillance and psychotic dedication

The most controversial aspects of *Homeland* remain focused around representations of psychological imbalance and mental illness, developed through complex depictions of government and terrorist organisations and motivating characters across all sides of our contemporary world. In a review section in *Cinema Journal* (54: 2015) edited by Diane Negra and Jorie Lagerwey, the varying contributors explain how Mathison perceives her public role as absolutely vital to national security and always prioritises it, over embracing more nurturing aspects of family life. In one contributing paper titled 'Women's bodies, Women and Mental Illness', Alex Bevan argues how the series dramatises 'physical representation' of the conflicts among aggressive US foreign policy, while also highlighting 'extensive domestic surveillance operations and the rhetoric of civil liberties' (Bevan 2015: 147). Such liberties which are under threat can also be extended to include climate injustice and the expectation of a clean and habitable environment. Yet surveillance per se, in spite of all the data analytical developments explored within this series, does not yield concrete or uncontested knowledge, much less clear and unambiguous explanations of the global problems of society. Again, like an ambiguous image extracted from a Rorschach test – albeit made up of digital co-ordinates – such ambiguity remains particularly prescient for this investigation. Recalling the high-tech scientific solutions proposed for dealing with climate change, which often involves an application of big data information sources, such high-tech and geo-engineering solutions are presented as a form of panacea for environmental and other threats, leaving no culpability for a variety of stakeholders, much less exposing any necessity for society to actively transition to a low carbon future. Quick and easy solutions are posited through the use of technological innovation, just like making a decision to carry out a drone attack, but without considering the full picture, much less adopting various checks and balances, including calling attention to various forms of moral hazard, or recognising and adopting the precautionary principle, etc. All these useful breaks on action and predictors of unintended consequences remain core considerations for environmental and sustainable practice and helps map out the possible negative consequences that should be assisted when making radical decisions that affect the safety of the whole planet.

Hollywood has of late become fixated with devices like drones, together with the power of big data-driven technology, both for their obvious aesthetic and surveillance appeal and for exploring the dramatic ethical concerns they throw up. See, for instance, Ethan Hawke's *Good Kill* (2014), or more recently the very successful cautionary tale *Eye in the Sky* (2016). Both of these features make dramatic play of the ethical dilemmas surrounding such robotic military operations, which mirror ongoing debates around using high tech and geo-engineering solutions to solve our growing political and climate change difficulties. The implications of global conflict are dramatically explored and re-imagined through such modes of surveillance. The enemy is only recorded at a distance on computer screens while being taken out, often resulting in what is euphemistically described as 'collateral damage' when civilians are killed. These broad-ranging ethical dilemmas are further pushed in a particular angle, on being instigated by an erstwhile nurturing female character, who is clearly codified as psychologically unstable. Yet, at the same time, Carrie (in contrast with Tony Soprano) embodies and represents a contemporary thin blue line between anarchy and civilisation, while trying to always do the right thing, including down the line helping to protect the planet.

'The Drone Queen': nurturing agenda and facing human responsibility and culpability

In the opening episode of Series 6 titled 'The Drone Queen', Mathison has been promoted to CIA Station Chief in Afghanistan for successful surveillance services carried out across previous series and has just received intelligence from Pakistan Station Chief Sandy Bachman (Corey Stoll) that a farmhouse in Pakistan is the current location of Haissam Haqqani, a highly sought-after terrorist. While there are some doubts regarding the provenance of this information, she asserts her power and authorises an air strike. Afterwards, the staff present their chief with a birthday cake, upon which she is proclaimed 'The Drone Queen'.

Responding to the narrative demand for conflict and drama, while simultaneously highlighting how everything is connected, the storyline develops by demonstrating that nothing is as simple or as clear-cut as it first seems. Making life-death decisions using the surveillance tools of audio-visual screens, is at least allegorically reminiscent of using a more nebulous series of (10) drawing blobs on a sheet of paper to help uncover a patient's psychological predisposition – both very different techniques for decoding reality remains faulty and open to contestation. Reports emerge from Pakistan that Haqqani was attending a wedding when the air strike hit, resulting in not just his death but also in the killing of over 40 civilians. Aayan Ibrahim (Suraj Sharma) a student who was injured in the explosion wakes up to find that his mother and sister were among those who perished. The rest of the series seeks to develop this fresh and unique perspective, as he plans his escape and seeks revenge. Meanwhile, mobile phone images of the resultant attack are put online. The footage goes viral, explicitly dramatising American complicity in such an atrocity and providing (clear) visual evidence of this reality.

In a world of fiction, such mistakes and catastrophes are inevitable and quickly turned towards questions of ethics, rather than having to deal with the immediate and messy on-the-ground consequences. Such juxtaposition can also be applied towards teasing out environmental justice issues and the rights and responsibility of different groupings across the world. Reminiscent of other long running successful television series like *24*, the capability of drones to supposedly remain in human control, while being co-opted to serve the greater common good is called into question. Naturally, of course, mistakes happen and audiences hopefully learn to appreciate the dangers of various unintended consequences, which can similarly occur when high-tech solutions are rushed through to help 'solve' global problems like climate change. Inferring that international global problems are solvable with high-tech solutions remains especially problematic with regards to climate change – whether this is measured at the scale of a planetary lack of water (or food etc.), much less the over-production of oil and dealing with the consequences of excessive CO_2 in the atmosphere. In this storyline, tragedy strikes when the operators or officer in command make bad decisions, or when terrorists secure control of the drones or the media publicity machine becoming a propaganda weapon, as extensively explored in series eight. Ethical issues are thereby focused on human behaviour patterns and decisions and how they are followed through. As unmanned AI controlled technology become more complex and more responsibility is ceded to these fighting machines, how much autonomy should the machines be allowed remains an abiding concern.[6]

Resolving this ethical dilemma similarly has major implications for dealing with climate change – essentially when and if various protocols for adopting large scale geo-engineering solutions, using the measuring capability of surveillance techniques, can be deployed effectively. Noting that such technical solutions have a tendency to go out of control is highlighted throughout this series. Yet these measures and devices are currently being actively considered in facing up to our growing CO_2 emissions crisis. Dealing with (environmental) ethical issues – be they population control, unstable democracies, surveillance, waste management, resource depletion, or any form of planetary limitations or threat – demands careful consideration and corroboration of scientific information that call attention to a multi-layered range of precautionary principles that *we* ought to be fully conscience of.

At the same time the media has a major role in addressing an interconnecting series of environmental and other global issues. One could suggest that these provocative series and story-lines at least sow the seeds for such debate by challenging audiences to take up more nuanced and ethical (political, cultural and environmental) standpoints, thereby potentially helping develop forms of critical political and environmental engagement. One might even hopefully suggest they can help produce new modes of critical media engagement and literacy. The series' function as a kind of a shop-window, dramatising complex personal, societal and political issues, while also drawing attention to the need for appreciating and appraising the range of topics addressed.

Environmental literacy: concluding remarks

Successful popular cultural artefacts like these provocative series alternate between transgressive and regressive polarities, especially with regards to how they represent humans and their inter-connecting political habitat and global environments. They are the nearest one can get within modern mass media to classic Greek, Jacobean or even Shakespearian dramatic engagement, while directly speaking to a new digital media culture and audience. This dramatic engagement assists audiences in getting an insight (both emotionally and cognitively) into these violent Mafia types, as well as exploring the mind-set of counter-intelligence officers and terrorist agents. Such creative and innovative televisual long-form series of dense psychological exploration, using elongated screen time, actively helps to tease out a complex range of political, ethical and environmental issues.

Media literacy in general, and environmental literacy in particular, remains the most fruitful approach around which to frame future audience investigations in this area, since it incorporates both an explanatory and a sometimes-provocative agenda for practice, all the while questioning that there is only one preferred form of narrative and aesthetic expression that can actively promote environmental learning. As Rust *et al.* (2013) have convincingly argued, *all* forms of popular media including television can be fruitfully co-opted to address environmental concerns.

Providing memorable enigmatic images such as the close-up of a dollar bill on a fridge door or that of a drone attack and the image of Carrie as a conflicted psychologically traumatised protagonist, serve both to suggest a new twenty-first century form of enigmatic Rorschach digital images for our audio-visual age. They might even help viewers (on the couch) unpack a range of often contradictory environmental and other messages played out across such narratives, while highlighting the ever present and destructive power of revenge. This is coupled with the need to engage with the Precautionary Principle for example, while deciding to carry out anti-terrorist and for that matter geo-engineering solutions to help solve our current environmental crisis.

Such populist dramas effectively recreate and evaluate the personal psychic needs and corruptive communal agency of virile men with guns like Tony Soprano, alongside recalling more unstable but equally potent female heroines like Carrie Mathison, who do what they can to help stabilise an often conflictual and revenge-driven (political) system. While not explicitly highlighting eco-gender concerns within this reading, such nurturing female agents co-incidentally embody a much greater psychological handicap, especially while appreciating an understanding of their motives and behavioural. At the same time, they do not in themselves promote a clearly codified environmental communications strategy to help resolve a growing number of global environmental disasters. Yet, surprisingly, such provocatively engaging protagonists do provide a useful frame of reference towards constructing an understanding of our fractured and environmentally compromised world. These popular generic protagonists are by any measure larger-than-life and sometimes even function outside the strict confines of their diegetic dramatic space. The contrasting world-views of gangsters and mafia, as against the treatment of international terrorists, speak to mass audience interests and preoccupations and often resonate long after their initial viewing. At least this is the rich semiotic potentiality of such high-quality televisual output, framing a broad range of complex and enigmatic scenarios, built upon two psychologically traumatised main characters. Simply figuring out and navigating the moral implications around how 'the end might justify the means' and dealing with multi-faceted environmentally-focused scenarios, while actively reading against the grain, remains an ongoing textual and ethical preoccupation for some viewers (and most especially communication scholars) to mull over and speak to.

Coda: an educational literacy agenda

In acquiring environmental knowledge through engagement with popular televisual series that are more than simply fleeting, it is essential that general audiences develop robust media and environmental literacy skills to help interrogate such drama. Concurrently, audio-visual media can play a significant role in facilitating participation in civic society and even promote new forms of activism, alongside aiding the development of environmental education across the public sphere, while at the same time helping to promote concern for environmental justice values for example that specifically cuts across inequalities at all levels. But to nail such hypothetical connections down and to understand this apparent slow process of audience engagement, much needs to be teased out, as environmental communications scholar Antonio Lopez (2013) asserts. In particular, far greater understanding of the potential power of (non-explicitly environmentally focused) fictional mediated texts is needed to explore how this might actively engage audiences, while promoting respect and care for the environment. Additionally, more needs to be known about how people's televisual watching might lead to environmental political participation. Such strategic investigation requires well-funded and cross-disciplinary comparative research over long time periods. At the same time, pedagogical practice and educational policy around such new forms of environmental literacy should be grounded in the experience of media use, learning, expression and civic participation across communities. Finally, this process should be developed and implemented through collaboration with academic stakeholders, schools and universities, the media industry and most importantly civic society across various regions (see Livingstone and Haddon 2011: 7).

Notes

1 Available online at: www.ejumpcut.org/archive/jc44.2001/aspasia/againstgrain1.html for a 're-reading' of such 'against the grain' strategies.
2 Incidentally it was based on an original Israeli series (*Hatufim/Prisoners of War*), which was created by Gideon Raff; *Homeland* is developed by Howard Gordan and Alex Gansa.
3 Lacanian film theorists call this process the oscillation between the unregulated desires of the Imaginary and the culturally sanctioned meaning-making of the Symbolic (McGowan and Kunkle 2004: xi–xxix).
4 See interview by Julian Hanich in *Necsus*. 2017. Available online at: https://necsus-ejms.org/vivian-sobchack-interview/.
5 Audience reception studies is however badly needed to map out such psychological trigger points and fictional identification connections.
6 Philosophically of course, questions can be asked about whether the machines ever have true autonomy in the first place (O'Mathuna 2015: 142).

References

Bellin, J. D. *Framing Monsters: Fantasy Film and Social Alienation*. Carbondale IL: Southern Illinois University Press. 2005.

Bevan, A. 'Women's Bodies: Women and Mental Illness' *Cinema Journal*, 54(4): 145–151. 2015.

Brereton, P. *Environmental Ethics and Film*. London: Routledge. 2016.

Brereton, P. *Hollywood Utopia: Ecology in Contemporary American Cinema*. Bristol UK: Intellect Press. 2005.

Dunleavy, T. *Television Drama: Form, Agency, Innovation*. Basingstoke UK: Palgrave. 2009.

Edgerton, G. *The Sopranos*. Detroit MI: Wayne State University Press. 2013.

Hall, S. 'Encoding and Decoding in the Television Discussion' Leicester University Papers UK. 1973. Available online at: http://epapers.bham.ac.uk/2962/1/Hall,_1973,_Encoding_and_Decoding_in_the_Television_Discourse.pdf.

Hiltner, K. (ed.) *Ecocriticism: The Essential Reader*. London: Routledge. 2015.

Lavery, D. (ed.) *This Thing of Ours: Investigating the Sopranos*. New York: Columbia University Press and Wallflower Press. 2002.

Livingstone, S. and Haddon, L. (eds) *EU Kids Online*. Nordicom. 2011. Available online at: www.nordicom.gu.se.

Lopez, A. R. *Greening the Media Literacy Ecosystem: Situating Media Literacy for Green Cultural Citizenship* PhD online from Prescott College in Sustainable Education. 2013.

McComas, K. and Shanahan, J. 'Telling Stories about Global Climate Change: Measuring the Impact of Narratives on Issue Cycles' *Communication Research*, 26(2): 30–59. 1999.

McGowan, T. and Kunkle, S. (eds) *Lacan and Contemporary Film*. New York: Other Press. 2004.

Modleski, T. *Feminism without Women: Culture and Criticism in a Postfeminist Age*. London: Routledge. 1991.

Morton, T. *Being Ecological*. Cambridge, MA: MIT Press. 2018.
Negra, D. and Lagerwey, J. (eds) *Cinema Journal*, 54(4): 126–131 [special issue on *Homeland*]. 2015.
O'Mathuna, D. 'Autonomous Fighting Machines: Narratives and Ethics'. In Hauskeller, M., Carbonell, C. D. and Philbeck, T. D. (eds) *The Palgrave Handbook of Posthumanism in Film and Television*. Basingstoke UK: Palgrave. 2015.
Parham, J. *Green Media and Popular Culture*. Basingstoke UK: Palgrave. 2016.
Pirkis, J, Blood, R. W., Francis, C. and McCallum K. 'On-Screen Portrayals of Mental Illness: Extent, Nature and Impacts' *Journal of Health Communication*: 523–541. 2006.
Polan, D. *The Sopranos*. Durham NC: Duke University. 2009.
Rust, S., Monani, S. and Cubitt, S. (eds) *Ecocinema, Theory and Practice*. London: Routledge. 2013.
Sandel, M. 'The Procedural Republic and the Unencumbered Self' *Political Theory*, 12: 81–96. 1984.
Short, S. *Cyborg Cinema and Contemporary Subjectivity*. Basingstoke UK: Palgrave. 2005.
Sontag, S. *Illness as Metaphor*. New York: Farrar, Straus and Giroux. 1978.
Weik von Mossner, A. *Affective Ecologies: Empathy, Emotion and Environmental Narrative*. Columbus: Ohio State University Press. 2017.

8 Netflix and emerging streaming networks

New forms of immersive and addictive narratives and characterisation

Overview

Across YouTube alone (as explored later in Chapter 10), not to mention streaming media like Netflix, there are over one billion uploads per minute. In 24 hours, this reaches 1.440 billion (Zehle in Maxwell *et al.* 2015: 74). All the while, the media industry has put on the technological 'automatic pilot', exclusively to pursue profits and build massive global audiences, as they oscillate between digital platforms and content providers. Such providers and platforms remain indifferent to causing environmental and social havoc, with so much e-waste being produced. Hence, global sustainability threats, deriving from new media products and ICT systems persist (see Kaapa 2018). One could even suggest that the social contract that held society and the economy together during the industrial era has apparently imploded, but unfortunately nothing concrete has emerged in its place.[1]

At a stretch, the so-called (clean) digital media's evolution online and into the cloud can encapsulate the allegorical back-story of *Breaking Bad*, where the main protagonist begins his descent into criminality by developing first as a 'cottage industry', producing illicit drugs out in the desert and polluting across wide open spaces, albeit on a small scale.[2] Over time, as reminiscent within the industrialisation process of development that has afflicted the planet with the call for growing economies-of-scale and supporting mass production, there is a similar transformational increased level of production of the illicit drug, ending up becoming a large-scale secret factory farm. This illicit enterprise and well-honed process is co-incidentally kept under-cover by the workings of a legitimate, but also environmentally unhealthy, fast-food chicken business.

This chapter seeks to conceptualise the uniqueness of streaming media and how it can provide new modes of convergence with televisual aesthetics, alongside exploring the ongoing broad appeal of long-play generic formats, both as a cultural practice and a particular style. Streaming has helped to tactically challenge the normative strategies of mass consumer and television culture with the capability for example of viewing programmes on hand-held devices and at any time of one's choosing, which in turn can make the physical act of viewing more unique. Furthermore, this chapter will explore how streaming might suggest modes of presenting and even critiquing long established environmental discourses.

For instance, in *Who Controls the Internet*, Goldsmith and Wu (2005) end their analysis with the onset of BitTorrent and communities of Internet users, who are able to circumvent state control but are also isolated and insulated from any kind of mainstream cultural impact (cited in Burroughs 2017: 61). The expediential rise of niche, instead of mass audiences, at first appears to make it difficult for a broad range of inter-generational audiences to coalesce around the global appeal of on-line media. Nevertheless, the phenomenal growth of cult and fan-based on-line mega-series, can be seen to speak to a new cohort of (mass) audience consumption, one not bound by the historical and material structures of conventional production, distribution or even reception.

Streaming basically operates as a contingent practice of 'making do' with the ever-changing modalities of production, distribution and consumption across contemporary consumer culture (Burroughs 2017: 64).[3] At the same time, recalling the concept of poaching, popularised by de Certeau (1984: 174), one could further claim that streamers can also be seen as 'travellers; they move across lands belonging to someone else, like nomads poaching across fields they did not write'.[4]

The long-established mass-entertainment industry, embodied by Hollywood and television studios, with their large back-catalogue of audio-visual content, can in turn solidify their revenue streams from marketing and re-distributing large archives and back-catalogues of audio-visual material they possess. This long-tail business model has most radically been re-positioned and augmented by global new media corporations, such as Netflix and others in the wings. All of which are scrambling to compete across the digital ether, while producing bundles of contemporary cult narratives that can be re-sold across various global market and through online outlets. This networked trans-national new media business model involves selling libraries of content back to the consumer, through as many possible outlets as possible. As is evident with the evolution of the music industry, by simply 'owning' a song as a licensed rental on one playing device such as an iPod that is legally non-transferrable to another device; a similar process can be replicated across other media formats. Some suggest that this revolutionary development in the music industry will most likely also rub off all new audio-visual media networks in the near future.

Consequently, control over content remains paramount. Media companies strive to lock down consumers in a timeless play that naturalises copyright law, making it seems like such new modalities and practices have existed forever and will continue to exist long into the future. This strategic logic is, however, a relatively new phenomenon, suggesting that works of art should belong to companies even beyond the life of the author or creator. But thanks to the panoptical practice of the Sonny Bono law, the life of copyright has been extended and effectively been frozen, in place of the public and society domain. Streaming of audio-visual content across a range of media formats similarly allows users to journey through popular culture and cut corners through retrofitting content to the contours of their own cultural imaginaries (Burroughs 2017: 67). Global media brands such as Netflix and Amazon are certainly direct benefactors from erstwhile forms of piracy, building on the informal niche-based media economy (Laboto and Thomas 2012), alongside adopting various other forms of contemporary pirate practice. All of which are becoming incorporated into mainstream business practices.

Nonetheless, the mainstream media industries at the same time still replicate a Fordist-style assembly line process, while the new Information Age is further characterised as mimicking a McDonaldised system of engagement, which is a long way away from the more benevolent and sustainable eco-system proposed by some media scholars. It resembles a complex network that remains fixated with moving people through its systems in predictable, incalculable ways, where individuals become more controlled, less empowered and somewhat dehumanised in the process. Such an analogy appears to contradict the so-called active agency of employees and consumers for that matter, who are fully in control of their media consumption and able to use and consume at their own convenience – all of which is greatly facilitated by the exigencies of streaming. 'Time spent on the internet or with digital media often has a light, playful, escapist dynamic' (Chayko 2017). Nonetheless, as evident in video games and to a lesser extent the streaming of cult television scenarios, play affords real benefits for audiences, in spite of the aforementioned insidious nature of new political economic tensions embedded within the industry. For instance, online television serial narratives can be viewed in any order or at any time, depending on the viewer's preferences. Pushing the notion of time shifting across scheduled television to a whole new level. Furthermore, this is an activity that speaks directly to audience's particular needs, as it is 'bounded from everyday life, separated from pressures and obligations' and 'freely chosen, non-instrumental, often absorbing and escapist' (Chayko 2017: 63).

The cult streamed televisual texts examined in this chapter – recalling the exploration of similar popular series like *Homeland* and *The Sopranos* discussed in Chapter 7 – are highly successful serialised formats, which appear in many ways far above the generic restrictions and aesthetic poverty of the soap format, yet continue to hook a new class of mass audience in a similar fashion. These new style series certainly emulate the much-disparaged melodramatic soap format that heretofore played often day-to-day, over many years, and enhanced the long-term involvement and enjoyment of their faithful audiences. More recently the technologically connected and networked fans are now able to condense such experiences over more intensive viewing periods and are also able to participate in their favourite media offerings via the internet, using the technical affordances of digital media. Audiences can easily extend their involvement and active engagement across such new media formats. For instance, at one extreme, with the aid of online media 'fans can readily discuss and critique programs, contribute to social media threads, blogs, and hashtags, and engage in communal discussions' (Chayko 2017:181).

Alternatively, this new form of access can be characterised as posing a Pandora's Box dilemma, recalling Barry Schwartz's identification of the paradox of choice. Essentially big mega-series, like those discussed in this and earlier chapters, demand many hours, weeks or even months of continuous and sustained engagement, all of which appear good reasons not to get involved in the first place. Yet audiences apparently find it very difficult to give up such hypnotic storylines, once locked in and hooked from the start.

However, as also suggested by some psychologists, too much choice can make us unhappy. Basically, the more choices one has, the harder and harder it becomes to feel happy. Consequently, there are concerns with the overuse of streaming through the internet, as well as across other mobile devices, creating a dependency on, or even an addiction to, such an ever-expanding range and proliferation of formats and reception devices. But, as aptly noted by communications scholar Dmitri Williams, 'if a person was reading novels, excessively, we'd be less likely to call that "addiction", because we value reading as culture' (cited in Chayko 2017: 192). So, for the purposes of this volume, we will keep an open mind by looking at how extensive consumption of such cult narratives with their sometimes-allusive pro-social environmental messages might generally be considered beneficial, rather than simply dismiss such forms of media consumption as just another example of mindless addiction.

Spending so much time in front of screens and sharing files, life stories, profiles, opinions, etc. through more interactive and systematic modes of interaction, it is implied that new generational audiences or users can more actively engage in, or become less healthy as a result, depending on the ethical position of the study. Online media research often focuses on prosumer activists with their pseudo-names, fake profiles, multiple identities and diversifying attitudes, as well as the endless efforts expended to re-position themselves vis-a-vis their favourite programmes. Such research can be further used to examine the potential learning received from the consumption of news for instance,[5] as well as in exploring fictional series across new communication technologies (Calik and Corbacioglu 2010: 4).[6] However, much of this expansive capability remains academic and often differs considerably from what is actually being consumed. By immediately focusing down on an exemplary televisual cult series, one can however hopefully appreciate the real benefits (as well as deficits) of such media outputs, in stimulating new forms of environmental engagement and even potentially promoting new forms of literacy.

How such engaging and hypnotic characterisation, explored in the following series, relate to nature and speak to an environmental sensibility – which might not appear obvious to the casual viewer – remains a primary preoccupation of this eco-textual analysis. Indeed, key questions addressed across all the narratives explored in this volume include:

- Where is (eco)environmental knowledge centred and how is it visualised?
- How effective are various modes of environmental education and even new forms of literacy being communicated within the narrative?

Situating *Breaking Bad* and promoting an environmental sense of place

Donal Carbough, in a forward to a 2017 edited volume '*Environmental Communication Pedagogy and Practice*', talks of the importance of place in communicating an environmental message and even transcending tensions between first and second wave modes of environmental criticism, as explored earlier in Chapter 2. Place certainly remains central to the cult series *Breaking Bad* and its prequel *Better Call Saul*, with their iconic American images of desert crossroads and other non-places where illicit drug deals go down. As Henry David Thoreau affirmed long ago, 'we are part and parcel of nature and it matters deeply where we dwell'. It matters deeply also where we imagine or think we are or express ourselves as being. A dominant form of this imaginative identification with place and habitat is spun out with (descriptive) words or audio-visual images. In *Breaking Bad* audiences are immersed into iconic images of hot deserts – recalling Western (fictional) territories and spaces – that are used to frame elemental conflicts between good and bad and at the same time highlight ethical concerns over 'doing the right thing' for environmental or other reasons.[7]

The main protagonist and chemist, Walter White, links toxic compounds to a place, diagnosing the ill-effects and re-imagining new forms of environmental harm in the process with his signature blue crystal meth drug. Note, this incandescent blue compound being produced is not a natural colour and both the process and product itself remain harmful and destructive to both nature and humankind. Hence, one could infer, the core drug-making storyline can be decoded as analogous to the constant harm humans are heaping on the earth, with their ongoing manufacturing of dangerous products.

On first witnessing this middle-aged and mild-mannered chemistry teacher (based in a High-School called J. P. Wynne), Walter White (Bryan Cranston) – famous for starring as the eponymous dad in the comedy television series *Malcolm in the Middle* – and his heavily pregnant wife Skyler (Anna Gunn), together with their cerebral palsy teenage son Walter Jr. (R. J. Mitte), for some strange reason, Somerset Maugham's (1915) semi-autobiographical *Of Human Bondage* comes to mind. In this classic coming-of-age tale from a bygone era, the main character Phillip Carey, who is an orphan and has a club foot, goes through awful personal trauma and anguish, only to apparently become trapped, when a girl he has only seen intermittently 'falls' pregnant. In the end however, Maugham shows how Carey realises that this apparent trap (which was one of the biggest imaginable in previous times), forces him to re-calibrate his apparently narrow field-of-vision for the future, and most surprisingly realises this was what he really wanted all along.

Walter White also becomes trapped, at first by his middle-class, albeit self-imposed poverty in America – where there is little protection, much less a financial security net – and then most significantly, becoming trapped as a random victim of terminal lung cancer. All of us desire what Anthony Giddens calls ontological security, 'the confidence or trust that the natural and social worlds are as they appear to be' (1984: 375, cited in Chayko 2017). There are of course so many complex psychological justifications used to explain the main protagonist's motivation, while trying to understand his fatalism, in his decent to violence and immorality. Most notably highlighted is his ever-deepening bitterness at not getting the requisite benefits from a pharma company he set up with two partners. Walter explains that 'Grey Matter Technologies' (the company he left behind for a miserly 5,000 dollar buy-out) is now worth 2.16 billion dollars. How does he know this – by looking it up every week. 'Walter is obsessed with regret. He looks back rather than looking forward. I sold my kid's birth-right for a few months' rent', he confesses. This is what seals his fate – 'He's too scared to take a real risk, so he seeks control instead' (Mattenson 2013: 2).

This pervasive sense of personal injustice has scarred him greatly and leads to his almost Shakespearian-like character flaw in developing a massive grudge, while having been apparently used and abused by the financial and employment system. Now, however, he will go to extreme lengths to take total control of his destiny. As a consequence, Walter takes on the most fantastical dark side of life, involving illicit drug making to help care for his growing family. But in the end, the series exposes how the main protagonist only really begins to 'live' (like so many Shakespearian tragic anti-heroes), as he grows to embrace the danger and excitement of such a Machiavellian alter-ego.

Most pointedly, as many reviewers have noted, Walter uses his illness and hard financial life generally as an excuse to 'break bad', knowing full well that when he's in remission, he'll have no excuse but to start being responsible. 'There's no other ending for this twisted story, because life requires accountability. We all owe each other an explanation, and nobody gets to do whatever the hell they want without consequences' (Mattenson 2013: 2). Shakespeare certainly knew this primary ethical weight of responsibility as a dramatic marker of character development, when he wrote his great tragedies, while for a new generational mass audience, such televisual series are equally resonant and even more engaging.

From allusions to Walt Whitman's romanticism and Tim Morton's HyperObect

Poetic allusions highlight a key romantic intertextual reference which helps to underpin the whole series, namely the romantic poetry volume *Leaves of Grass* by Walt Whitman. This reference affords a big reveal very late in the series for Hank (Dean Norris), who finally realises that the almost mythic drug-maker he is chasing and going by the name of Heisenberg, is in fact his brother-in-law Water White. This shocking realisation is sparked by remembering back to when Walt was confronted by him at a dinner party and quickly if jokingly explained a piece of evidence with the initials 'WW', as not referring to him, but rather the great romantic poet Walt Whitman.

Now at this late stage in the story, Walter has finally decided to throw in the towel, having physically seen the spoils of so much drug money stored by his wife in a locker room space and literally not knowing what to do with it. The family is subsequently having a chilled-out meal together with their ever-present in-laws Hank and Betsy (Maria Schrader) near their outdoor pool. All at last appears on an even keel for their family. But suddenly, the penny drops for Hank, as the investigative DEA officer slowly realises the truth, which has been under his nose all this time. On the toilet, he examines a copy of *Leaves of Grass* left lying around, while looking at the inscription which reads 'WW'. At last the connection is made and the secret is out.

Audiences, of course, are used to acquiring such knowledge well in advance of the protagonist in a story and are always primed for such secrets to be revealed, as the longer it is delayed the more excitement and gratification is engendered.[8] Furthermore, using this iconic romantic and nature poet – both as a narrative device and an important reference point – betrays a clear indication of the series' environmental theme and even I would suggest its growing preoccupation, if only through hinting at Walt's secret interior life and his apparent love of nature poetry – knowledge, of course, audiences don't otherwise have access to. This is not the theatrical convention of a soliloquy, or the modern mode of psychological therapy, recalling the benefit of the 'talking cure', as dramatised in *The Sopranos*, for example. At last some tangible evidence of Walter's environmental sensibility and his deep respect for at least the allegorical power of words are revealed. One presumes that Walt actively rereads and enjoys nature poetry, while secure in the toilet – the only safe space where he does not have to wear a mask. This iconic eco-poetry book, gifted to him by his drug-making partner Jesse (Aaron Paul) – signifying their surrogate father/son bond, together with their master/apprentice relationship – might at a further stretch serve as a worm-hole link and jumping off point (recalling Morton's notion of the hyperobject[9]) to a potentially deep form of environmental engagement and connectivity. Allegorically, at least, the power of romantic and nature poetry highlights the main protagonist's hidden pro-social and environmental sensibility and even suggests a more appealing alter-ego for audiences to relate to.

This revelation and environmental cross-connection is also signalled like a piece of a puzzle, as viewers remember back through the various series and recall the recurring incidence of animals and especially insects – from bees to other forms of insects – all serving to embody and dramatise a vast desert and brutal habitat, which becomes the site of so much nefarious activity. Through all of this, 'WW' maintains his love (or one would at least like to think so) for the 'therapeutic power of poetry', to help soothe his troubled soul.[10] Such an eco-character analysis further recalls the power of biophilia and the long-term benefits of first wave environmental connectivity.

Re-calibrating the artistic critique and discussion of so-called second wave environmentalism in Chapter 2, which does not demand direct exposition of environmental issues to be resonant in the text, this allusive narrative trajectory uncovered in *Breaking Bad* is not necessarily about celebrating nature or the environment, but rather, if at some remove, calling attention to such concerns.

Nonetheless, such narrative insights serve to re-state the primary significance of nature as a sacred place.[11]

Furthermore, this set of binary nature oppositions is set up as a backdrop to the ever-expanding diegesis of the storyline, helping to flesh out the main protagonist's growing and apparently (un)ethical mode of environmental behaviour. For instance, throughout the series, audiences frequently observe the business of rolling out the barrels to dispose of physical human remains as evidence, while at the same time they are also used to hide the illicit drugs during transportation. A well-oiled machine, involving nefarious drug production and distribution, pushes waste reduction and disposal to its ultimate extreme – recalling the preoccupation of earlier gangster series like *The Sopranos*, who also ran a very unecological waste disposal business.

Other nature signifiers (if not necessarily developed or extended as hyperobjects) that come to mind throughout the series include a memory when his brother-in-law is at a very low point of his professional career as a narcotics investigator and having numerous suspects assassinated at the same time in jail, just before they can talk and finger the big drug leader. Hank reminisces with Walt around wanting a 'simpler life' and recalls his first job, marking trees for cutting in a forest. It was hard work but honest and better than the madness of 'catching evil characters'. Meanwhile, Walt looks on bemused or maybe even intuitively connecting with such biophilic reminiscence. Again, such anecdotal revelations speak directly to the vicarious viewers also looking on and trying to understand (if not necessarily empathise with) the protagonists differing mindsets. The hunter/hunted trope, of course, has remained a major preoccupation across so many genres in both film and literature. Similarly, we can also recall the 'thin blue line' of police security striving to protect society from the irrational madness and anarchy of the criminal underworld, where the ends always justify the means.[12] Such complex narratives tropes and character developments encourage audience's complicit pleasures in engaging with such conflicting amoral behaviour, rather than striving to 'do the right thing' for their fellow human beings, much less for their environment. All the while, Walt appears at sea, without any ethical compass or set of values, spiralling deeper into a negative psychological and malevolent vortex. Unfortunately, such a strong-willed, dramatically conceived (even Shakespearean-like) protagonist is seldom satisfied with good, wholesome and environmentally beneficial work, much less looking out for the long-term betterment of communities and society generally.[13]

Deep green environmental values and the ticking clock of poor health: cancer!

Together with Walt Whitman, one could most definitely use the analogy of cancer, which the main protagonist is presented with from the start of the series, as symbolically echoing the world's current unwillingness to recognise, if not face up to, the spectre of climate change. As this deadly cancer is eating into his fictional body, Walt initially decides to ignore it and hatches a plan to provide for his family, by following probably the most unsustainable – albeit criminal – form of activity that can nonetheless at the same time produce unimaginable material wealth. So much so, in fact, that, in the end, as mentioned earlier, it takes his wife Skyler and the physical evidence of his money accumulation to finally realise that enough is enough. Such protagonists are primarily driven apparently by the acquisitive and so-called selfish gene and seek to continuously acquire more and more material wealth. Acquiring excessive wealth is always justified as a reason to protect and provide for their loved ones, by using any means available.

How to reverse this trajectory of ceaseless acquisition and greed coincidentally underscores a core dilemma also for environmental rectitude. Striving to support alternative storylines that seek to reverse the dominant selfish drive of human behaviour and help move away from the individual-centred and global treadmill of continuous economic and material capitalist growth remains constantly in the background. All of which involves a paradigm shift in contemporary norms that few can even imagine, much less activate. Consequently, as this volume suggests, audiences need the active support of powerful visualising models, and most importantly the assistance of deep allegorical storytelling triggers, to help re-imagine the effects of climate change while at the same time constantly revealing what crass and illicit wealth looks like. More importantly, the narrator provokes some level of acknowledgement around the necessary creation of a cautionary tale or even suggesting a 'teacherly text' designed to help signal how humanity must move away from the edge – almost reciting the religious parable. Such cautionary tales outline how material wealth is of little use in the long term, if our physical and/or mental health is not being nurtured, or is working only to promote individual rather than familial and most importantly communal solidarity and well-being.[14]

Yet moving beyond allegory, as Paul G. Harris (2013) and others affirm, there remains genuine disagreement about how best to perceive what is wrong with climate politics and most importantly how to fix it – unlike cancer or other bodily ailments – while 'continuing to do what we are doing now is certainly no solution. That would make the problem worse.' Such evocative and globally successful cautionary tales like *Breaking Bad* can somewhat surprisingly and counter-intuitively help make a start, at least, in calling attention to such mind-blowing interconnecting issues – even if only allegorically, or tangentially and against the grain, as compared with the actual on-the-ground political and environmental sources and possible solutions to this ongoing dilemma.

But even more provocative, from a specifically environmental perspective (if also thankfully a more humorous text), is the introduction of a low-life lawyer James (Jimmy) McGill, lately known as Saul Goodman (Bob Odenkirk) who becomes a revelation and helps to bring some humanity to *Breaking Bad*. In turn, his stand-out character became so popular, he spawned a very successful series in his own right, titled *Better Call Saul* which we now turn to.

***Better Call Saul*: new modes of professional practice that can address planetary problems**

The spin-off series focuses on 'slippery' Jimmy (Saul) and his older brother Chuck (Michael McKean) who is the 'real' lawyer (see his similar role in *Homeland*) and who has always had to bail out his younger brother from trouble. The narrative arc follows Jimmy's (pre-*Breaking Bad*) history and motivation and love for his legal partner Kim Wexler (Rhea Seehorn), while fighting against his old boss Howard Hamlin (Patrick Fabian) and all the while dealing with his blood brother, who has serious mental health issues. Eventually, the newly-reformed Jimmy McGill gets a job with Chuck in his law firm, which the older-brother co-owns with his partner Howard. Jimmy uses his time wisely to get an online law degree, ostensibly to make his big brother proud. But, of course, sibling rivalry runs deep. We learn, from flashbacks of their childhood, how the young Jimmy stole from his father's shop and eventually – at least according to his older brother – drove the family into bankruptcy. Apparently, an erstwhile huckster looking for money from his father taught the young boy a valuable lesson that he never forgot. 'There are only two types of people: sheep and hawks: you need to decide which one you are.' In taking the huckster's message to heart, Jimmy becomes a master at reading people and uncovering how to carry out the ultimate con-job.[15]

Alternatively, always striving to 'do the right thing' and following the black and white exigencies of the law encapsulates the DNA of his older brother, Chuck, who is an upstanding citizen and always does things by the book. Stereotypically, Chuck has predetermined core ethical values and is not slippery like his brother. But, as dramatised as far back as Milton's epic *Paradise Lost*, audiences and humanity generally prefer a 'fallen angel' who does not necessarily do things by the book. Such characters appear more interesting, simply because they are flawed and thereby more human like us. Further memories and flashbacks by Chuck confirm this truism – recalling, for instance, a scene with both brothers at their mother's dying bedside. When Jimmy goes off to get some food, Chuck witnesses their dying mother's last words in calling for Jimmy, rather than her more faithful son. Chuck is so annoyed, he never reveals this incident to his brother. Meanwhile, Jimmy has no qualms in telling stories against himself, no matter what the consequences, as he discovers to his cost. Nonetheless I will argue, Chuck embodies the necessary values and attributes to support the dramatic move to achieving a low carbon future.

Chuck: allegorical embodiment of frugality (a cautionary allegory on the dangers of fossil fuel energy!)

Reminiscent of the female protagonist Carol White (Julianne Moore) in Todd Haynes' *Safe* (1995), which portrays a 'normal LA housewife whose life is shattered when she develops an allegory to the twentieth century: the chemicals that are everywhere in food, fabrics and fragrances which led to breathlessness and nausea' (Brereton 2005: 174). White ends up in a hippie-type igloo trying to save herself. Similar to Walter's cancer in *Breaking Bad*, Chuck's strange illness functions more directly and on a bigger environmental scale for the twenty-first century. The older brother is traumatised by all forms of electric current, be it from so-called smart mobile phones or everyday technology, such as electric lights and kitchen utensils.

Incidentally, only much later in the series do audiences discover the trigger to his illness, which suddenly occurred when his marriage to his perfectionist and classically trained musician Rebecca broke up. Hiding his psychological illness from his ex-wife, until the final big show-down court case between the two brothers, Chuck names his illness 'Electric Magnetism Hyper Sensitivity'. A disability not recognised of course by the medical profession, who naturally want to carry out tests to discover what is empirically as well as clinically out of synch in the patient. Chuck justifies how it takes time to explain to the general public such a new medical condition, as was the case with AIDS and for that matter other contemporary environmental conditions, including the fictional one Carol White railed against in *Safe*. Chuck is stigmatised for his condition, while suffering greatly. Some naturally regard his illness as a form of mental delusion, displaying no physical causes and therefore not easily categorised and hence dismissible. At several stages across the series, Jimmy could have had his brother committed indefinitely, but because of unconditional love and family loyalty he never even considered such a plan of action.

Allegorically, Chuck's condition certainly echoes and provokes an engaging environmental metaphor, while affording a useful creative imaginary and allegory around the inherent dangers of fossil fuel energy and its long-term consequences for climate change. With so much CO_2 and e-waste being produced by so many mobile phones and electric gadgets of all kinds literally 'killing the planet', his resulting albeit pathological coping strategies serves as a potent, if somewhat exaggerated, environmental cautionary tale. While of course audiences, even subconsciously, may not make such a direct, or even an allegorical correlation to environmentalism per se, one can, nevertheless, suggest that the series puts such issues directly into the public consciousness. Consequently, the Netflix series serves one of the roles of environmental education, calling attention to such (tenuous, even against the grain) connections that can in turn be highlighted and illustrated as part of a broad environmental literacy agenda.

Yet, at the same time, as a successful senior lawyer in a big firm who digests macro conflicts as well as adjudicating on micro legal cases, Chuck has the intellectual acumen and tenacity to uncover truth and deception in all its guises. No more so than in a case involving OAPs who are residing in a franchised nursing home, while being secretly fiddled out of their savings by being systematically over-charged for services rendered. Jimmy first uncovered the illicit dealings and then goes to great extremes to discover clues needed to prove the case in a court of law. He literally fishes through waste bins to find hard evidence, which of course is more usually the provenance of investigative journalists and reporters. But only Chuck has the tenacity to stay up all night, teasing out reams of shredded paper and reconstructing them back together to finally uncover concrete clues – thereby being able to build a strong case against the nursing home management team. Such low-tech teasing out of jumbled up and shredded bits of paper acts as a potent visual counterpoint, which contrasts with the ease of analysis afforded by computerised big data-sets – a digital process which is illustrated, for example, by a reading of *Homeland* in the previous chapter and carried out almost instantaneously by powerfully conceived computer technology at the press of a button.

Pragmatically, it will require legal (environmental) experts of the calibre of Chuck to successfully fight off the dominant fossil fuel companies – like the tobacco companies in earlier times – and to patiently and forensically demonstrate their illicit abuse of CO_2 resources and thereby highlight their direct effects in destroying the planet.[16] Furthermore, calling on business, insurance, security and legal sanctions are all part of the most tangible mainstream approach to instigate adaptation and global change within the environmental landscape. Such disciplines and strategies, unfortunately, are not always recognised within Humanities and Social Science – actively using existing professions, tools and protocols to fight against climate change. Hence such provocative and allegorical filmic narratives and character tropes – even if not always ostensibly speaking directly to environmental issues – can be persuasive in illustrating both established and new ways of actively engaging with such complex environmental problems.

In coping with, while also at the same time calling attention to, his affliction, Chuck insists that no electrical devices are allowed into his house. For his rare trips to the office, the company has to go on 'lock-down', confiscating all staff mobile phones and turning out all the lights. At a stretch, could such a radical protocol serve as a cautionary exemplar of an extremely purist form of transitionary post-carbon future society? At a further stretch, it might even infer a radical but very un-appetising model of frugal (pre-modern) living, before electricity literally transformed our world.[17]

Nonetheless, by constantly treating Chuck as disabled, if not also psychotic, the situation remains at the edge of normality and thereby becomes easy to dismiss out of hand. Yet, audiences might almost, through osmosis, learn to understand and appreciate the very strange and difficult situations involved, while trying to imagine a post-carbon energy future, even if at some remove. This is reminiscent, though probably at a much lower scale, of how new-generational youth might experience a power outage for the first time in their lives and in having to cope without the everyday benefits of electricity. How can *we* possibly re-imagine a low carbon environmental world, if not by initially acknowledging, if not necessarily embracing, the frugal extremes of a pre-electrical era and at the same time creatively envisaging a pre-modern and pre-industrial environmental landscape. Engaging with social and technological history as a process and model of human engagement is helpful for a renewed critical engagement with politics, culture and economics, especially as framed towards addressing all our technological and environmentally predetermined futures.

Chuck's fatal flaw however (like Hank's to a lesser extent in *Breaking Bad*) is his sanctimonious belief in his own innate powers and truth-telling capability. This is reminiscent of a religious demagogue who always believes they have a monopoly on the full environmental truth (see, for instance, an analysis of the religious protagonists in *The Mosquito Coast* (1986) discussed in Brereton 2005), which corresponds with the drive of some egotistical climate scientists, and even fundamentalist environmental activists, who espouse the right to define the only way to address solutions to climate change through a restrictive and one-directional appreciation and understanding of 'settled science'. While Chuck most certainly knows the limitations of his brother Jimmy, unfortunately he pushes him too far. Affirming his own stoical probity and self-righteousness in appreciating the absolute truth of the legal system to speak over all other discourses, Chuck is unfortunately speaking to only one aspect of truth-making and problem solving. Later, when not accepted for his actions, even by his legal peers as the only meta-truth, he is unable to cope. Jimmy, on the other hand, is more human and has to be taught such environmental appreciation.

Fly-tipping, education and confronting an environmental allegory

At a meta-allegorical stretch, Jimmy embodies a more mainstream view of the foibles embedded within human society, which is unconsciously ignoring the problem, or at least not actively dealing with environmental issues and hoping to get away from their long-term consequences, as they slip and slide around the truth. He encapsulates most of the general populace, who constantly espouse the path of least resistance. One of the few times Jimmy has to directly confront the consequences of his actions occurs later, after the brothers' trial, on being assigned community service in reparation for his wrongdoing.

As part of his assigned community service, the ever-cocky and new legal expert Jimmy ends up literally picking up all the refuge and waste randomly discarded from the cars travelling at speed on the highway above.[18] From soiled nappies and other discarded bits of waste, audiences are shown the full horrors of fly-tipping. But one wonders if such under-valued work teaches any environmental, much less ethical, lesson to such flawed self-centred individuals like Jimmy/Saul, who have coincidentally avoided going to jail (as also witnessed across other series like *The Sopranos* or Walter in *Breaking Bad*) by taking on such community service. Not likely, I suspect! But hopefully the lesson is not fully lost on the general audience.[19]

Only later, after his girlfriend nearly dies in a car accident, having stayed up too many late nights to get a legal job completed, does Jimmy consciously and carefully pick out wads of legal paper documents spilled all over the highway edge. All of this activity affords further evidence of the ongoing pollution of the landscape, which can at least metaphorically be extended to various other levels of waste disposal across modern society as illustrated in other stories.

At a macro level, one might, for instance, share Val Plumwood's argument around the need to overcome 'the ecological crisis of reason' – in this case embodied by Chuck's persona – where decisions are made seemingly on the basis of dominant technical-rational evidence that often appears to override emotional, experiential, human-environmental and dialogical relationships (Brown 2017: 221). But if the alternative is abusing the law simply for selfish gain, as embodied by 'Slippery Jimmy' – even if communities get the 'right result' one must wonder what might the unintended cost involve? Ideally, one would hope that the big picture and purist legal and rational mind of Chuck, could be combined with the pragmatic yet endearingly selfish character of Jimmy, who knows how to deal with people and can use political exigencies to convincingly address wicked problems like climate change.

Environmental shifts in perceptions: how to make frugality exciting

Hypothetically, at least, one further wonders, if like a puritan priestly-character, Chuck represent the 'real fatalistic truth' of modern life and also serves as an allegorical exemplar, with his life story constituting a dark cautionary tale around what might happen if modern society does not take control of dealing with the truth (with so much 'fake news' around), including the logical outcomes of ignoring or mis-handling facing up to the scourge of climate change. Chuck consistently believes in the law as being mankind's greatest achievement, recalling how no matter who you are, your actions have consequences. Such laudable precepts speak to a robust meta-narrative which all citizens ought to, de facto, subscribe to. An environmental system of legal checks and balances, built on rules that have to be obeyed no matter what the cost, this remains a foundational principle of modern democracies. Unfortunately, as we know throughout history and across geographical regions, such rational and legal sanctions and ethical coherence is often fractured and certainly open to abuse and contestation. Hypothetically, however, if Chuck was in control of the world, the global climate change dilemma would probably be solved – but at what cost? At least the hope remains that varying and complementary disciplines coupled with the agency of contrasting personalities and ways of creating change – as embodied by blood brothers Chuck and Jimmy – can combine in some more constructive way to create effective long-term environmental solutions to such wicked global problems and in turn help push society along a more sustainable pathway.

Incidentally, Shi posits that the United States has had a cyclical relationship with frugal or simple living, as embodied by Chuck: from colonial era Puritans, to Quakers, and even modern back-to-the-land and most recently the hippie movements (Boucher 2017: 224). Environmentalists across the world, not surprisingly, remain more prevalent among higher income cohorts, who are coincidentally and paradoxically the most polluting members of society. We can certainly perceive this through the character of Jimmy McGill (noting his quick-witted Irish background and stereotypical easy-going nature) and who would not even pretend to espouse a deep form of environmental ethics as explored above. Alternatively, Chuck appears more WASPish and frugal in his high-class, but austere, consumption patterns.[20] Chuck's whole behavioural and moral compass is totally defined by the law, which projects a defining meta-narrative that has to be obeyed by all and therefore, I repeat, can more effectively speak to other complex issues like climate change. At the same time the law remains only one of the many strictures or performative roles Jimmy (and one suspects a majority of citizens) strive to adopt.

While taking into account his disability, Chuck certainly represents an environmentally sensitive or frugal disposition that speaks to a radical yet long-established logic that might meet the perceived cultural stereotype of a pure climate change activist – one who is willing to rise above the usual 'business as usual' modalities and recognise the massive intellectual and behavioural transformation that is required to address what must be done by Western (and global) society, in facing up to the challenges of climate change (see for instance Paul Schrader's 2018 dark climate change film *First Reformed*). His persona has the intellectual and personal tenacity to imagine taking on the macro- and micro-system of systemic change and the vision to address the demands for supportive green regulation, alongside other legal strictures necessary for environmental transformation.[21]

Frugality, as an opaque rather than a restrictive notion designed to limit human behaviour, in accordance with prescribed restrictive conditions, has to be re-imagined and distributed into a taxonomy of logics. It can either be constrained or simply dismissed as habitual or nostalgic, while contradicting one's economic class status. Yet it cannot, of course, be the only solution. Frugality, as one critic asserts 'may even reduce one's carbon footprint, but I would question such a finding, especially among the affluent: I question the environmental relief that would follow from such a seemingly individualised and internal project' (Boucher 2017: 231). Certainly, as Boucher and others suggest, frugality is no panacea to address climate change. But how these conflicting attributes play out through the mildly eccentric and electricity phobia of Chuck functions as a dramatic case in point. To put it somewhat crudely, this reading wonders out loud how a flawed and mentally unstable personality (recalling other characters discussed in the previous chapter like Carrie in *Homeland* or Tony in *The Sopranos*) might be re-appropriated and re-constituted as a touchpoint for more progressive environmental representational agency? Much pedagogical engagement and textual analysis is needed to tease out the potentiality and range of such provocative exemplary fictional narratives, while at the same time fleshing out such three-dimensional fictional characterisation.

All the while, various forms of narrative and character analysis can be applied as a means of crystallising arguments and environmental assumptions. Furthermore, storytelling can be usefully used as a proactive way of understanding, communicating and influencing others, which can become almost infinite in its long-term effects. Environmental scholars must, however, reflect on some of the challenges and possibilities for continuing to develop such online televised 'stories' as data sources for modes of inquiry, and hopefully as creative paths towards active social and environmental engagement. Most importantly, such images and storylines afford new creative images to engage with a growing range of environmental dilemmas, as citizens face global insecurity in the modern world. Remembering our understanding of the planet is always based on emotion and personal belief, as well as sometimes contradicting physical and measurable data (Moezzi *et al.* 2017: 1). If nothing else, such provocative characterisations and storylines open up fruitful environmental avenues for ongoing discussion.

Concluding remarks: narrative tropes that drive these streaming TV series

Bruner, in a useful study, argues that human beings understand the world in two distinct ways, defined as the 'paradigmatic mode' of thought and the 'narrative mode' of thought. The former exemplified by scientists and logicians, which would also include 'straight by the book lawyers' like Chuck in *Better Call Saul*, who seek to understand and determine cause-and-effect relationships and develop tightly reasoned analyses, logical proofs and empirical observations (Brown 2017: 216). While, alternatively, the more 'flawed humans' creative and narrative mode, as embodied by many of the main protagonists in mainstream Hollywood tales and across such Netflix series – characters like Jimmy and, of course, Walter – seek to re-imagine the world to conform with their unique mindsets, Jimmy McGill simply wants to make as much money as he can, fulfilling the usual American Dream ethos, like Walter in *Breaking Bad* if at a more extreme level – neither character wants to be taken for a sucker.

Narrative theorists believe that stories and storytelling remain a potentially important device in helping people from different disciplines and different cultural domains – to better understand the world and each other – while also facilitating and working on applied environmental problems, including imaginative story-worlds that walk outside normal constraints (Moezzi *et al.* 2017: 2). Many scholars suggest that a story must relate to past events and must have a point. By any stretch, such psychological and biographically rich narratives like *Breaking Bad* or *Better Call Saul* are driven by a complex network of associations, interactions and biographical trigger points, which both inspire and define the contours of flawed human behaviour that speak to audiences generally.

McCloud further asserts that one of the central functions of a story ark is that it offers 'guidelines', 'a script', or suggests ways that one should behave in certain circumstances. According to Murray, the prime function of narrative is its ability to bring order to disorder, which no doubt is badly needed in facing up to the prospect of environmental disaster. Alternatively, Becker suggests that the use of narrative is particularly pronounced, when people try to make sense of 'chaotic' events and disruptions to routine, such as personal, financial problems, health and environmental challenges (cited in Brown 2017: 218).

Better Call Saul probably attempted to do too much within its televisual diegesis, with too many non-interconnecting storylines trying to mesh together. Nonetheless, the series remains fascinating on so many levels and speaks to real-life issues from the standpoint of legal strictures to address problems, including environmental ones. This eco-reading illustrates how Chuck and Jimmy remain the yin and yang of alternative strategies in dealing with environmental problems (and general life, for that matter). Meanwhile, looking back to *Breaking Bad*, at one level the series remains an unconventional re-working of the Bonnie and Clyde myth, as every known stereotype is turned on its head. What strange bed-fellows this new form of networked television is developing, which most certainly has widened the creative imaginary of narrative templates across the landscape of new audio-visual media. Such an innovative series at least covertly speaks to complex issues like environmental disruption, while finally recognising the shocking implications and consequences of the most dangerous manifestation of global disequilibrium encapsulated in climate change.

Notes

1 By allowing poisonous chemicals, metals and radiation, ICT firms continue to pollute vital resources that affect humanity's habitat heritage: the environment (Zehle in Maxwell *et al.* 2015: 82).
2 See the interesting reading in McDonald and Smith-Rowsey's *The Netflix Effect* which affirms that the end of *Breaking Bad* can be contrasted with the rise of Netflix: 'the hero was defeated after having vanquished a slew of economic adversaries. In a way, when he fails to become a mogul with mass distribution, Netflix has succeeded, primarily by dominating the video streaming market' (2016: 196).
3 Streaming is a polysemic practice which is not completely autonomous, but at the same time not completely vulnerable and dependent, as a pre-defined spatially-based audience segmentation.
4 Altman for instance argues that every 'new' media technology goes through a process of negotiation where the 'definition of a representational technology is *both historically and socially contingent*' (Altman 2004: 16). Just as there is a 'rhetoric of walking' (de Certeau 1984: 100) we might think of a 'style of use' (de Certeau 1984: 100), for streamers traversing the geographies of nation-states and the lands of copyright holders throughout the internet (Burroughs 2017: 56).
5 Virtual networks create diversity in society, so reflexive society create reflexive identities. The risk of stereotyping also impacts, and there will be many identities resembling each other in the same community (Gunduz and Pembecioglu 2014: 39, cited in Gunduz 2017: 86).
6 There are few 'no exit' relationships in cyberspace, and the term 'surfing the net' remains an apt descriptor of the depth of much online activity. In communication processes, internet use allows people to establish interaction by ignoring voice, image and biological and social gender for the first time. Users can establish an identity for themselves by creating any profile that they want and can freely share any information (Boyd and Ellison 2012: 1, cited in Gunduz 2017: 88).
7 For instance, recalling the fallout from murdering a boy on his bike who is simply out in the desert landscape 'exploring nature' and thereby conforming with proactive environmental learning protocols, tells us that all young people should be engaged in fruitful biophiliac practices, by exploring local habitats and landscape. See also an analysis of Pokémon Go in the following chapter.

8 Such revelations play off knowledge and narrative secrets, which characterise much of the differed pleasure of soaps and long-play televisual series.

9 See Tim Morton's book *The Ecological Thought* which is echoed in his latest study *Being Ecological* (2018), where he re-defines the hyperobject,

> as a thing so vast in both temporal and spatial terms that we can only see slices of it at a time: hyperobjects come in and out of phase with human time, they end up 'contaminating'everything, if we find ourselves inside them (I call this phenomenon *viscosity*). Imagine *all the plastic bags in existence at all*: all of them, all that will ever exist, everywhere. This heap of plastic bags is a hyperobject: it's an entity that is massively distributed in space and time in such a way that you obviously can only access small slices of it at a time, and in such a way that obviously transcends merely human access modes and scales.
>
> (Morton 2018: 77)

10 Earlier in the desert, Walter also reads Percy Shelly's poem *Ozymandias*, while the *mise-en-scene* represents and inhabits a barren New Mexican desert – noting the poem is also set in such an erstwhile barren space. (See, in particular, Episode 14, Series 5, Ozymandias – From a poem by Shelly: 'My name is Ozymandias, King of Kings./ Look on my works, ye Mighty and despair.')

11 Place does not often 'speak' to new generations – first hand – so we need to 'learn' how to engage with the help of dramatic narratives to help tease out how these teach us how to engage environmentally with place.

12 Incidentally this trope is repeated at a key dramatic moment by Chuck to his brother Jimmy and the role and importance of an objective and universally agreed legal system in the related *Better Call Saul* discussed presently.

13 As the American Discovery channel, suggested by a study by Murphy (2017), implies, character-driven narratives remain the best way to speak to audiences and even promote new forms of environmental engagement and literacy.

14 This strategy and appeal is also attempted in anti-business movies like *The Wolf of Wall Street* and other narratives, but are always in danger of being complicit in what they are railing against. A big ask by any measure for a populist story-line and cautionary tale, which has been traditionally the provenance of religious allegories?

15 We later see his over-weight buddy from the old-days, as they nostalgically re-do one of their old tricks. Basically, the friend lies in the gutter as if dead, but not before pronouncing that he was never happier than now, in doing what he did best.

16 This is reminiscent of the true story environmental legal fight celebrated by the film *Erin Brockovich* (2000).

17 See for instance an analysis of the Amish community who espouse such pre-modern modes of technology as dramatised in films like *Witness* (reviewed in Brereton 2005).

18 Note, Community and Tidy Town groups do this form of environmental work freely and for the good of everyone. The real scandal is that it has to be mandated to individuals as part of community service to help keep their environment free from waste.

19 Eventually, of course, Jimmy learns to abuse the system, using his litigation skills in securing release of a fellow inmate who has a drug deal going down, by making the argument that he will use personal litigation against the officer involved if he does not allow his 'friend' to go 'home to his family' as requested – all such para-legal services for a sizable fee of 700 dollars.

20 He can be decoded using a 'Bourdieu approach', meaning that one can consider how people like him are socialised and positioned for earning cultural and economic profits in certain fields of struggle – namely the long-established legal profession (Boucher 2017: 225).

21 However, for instance when climate change activists themselves hypocritically making fun of each other for being frugal, it is difficult to see how the notion of frugality as a potential fruitful attribute, much less going frugal in one's lifestyle and

embracing the radical departure of de-growth, or the more acceptable notion of a 'circular economy', might be more positively accepted and encouraged across intellectual debates in society. See, for instance, the ongoing mass media send-up of vegans mentioned in the food chapter, much less any form of back-to-nature bourgeois living, reported in the news. Pro-environmentalists are unfortunately not perceived as exciting, much less as being 'cutting edge' or sexy!

References

Altman, R. *Silent Film Sound*. New York: Columbia University Press. 2004.

Boucher, J. L. 'The Logics of Frugality: Reproducing Tastes of Necessity among Affluent Climate Change Activists' *Energy Research and Social Science*, 31: 223–232. 2017.

Brereton, P. *Hollywood Utopia: Ecology in Contemporary American Cinema*. Bristol UK: Intellect Press. 2005.

Brodesco, A. 'Heisenberg: Epistemological Implications of a Criminal Pseudonym'. In David P. Pierson (ed.) *Breaking Bad: Critical Essays on the Context, Politics, Style, and Reception of the Television Series*, pp. 53–72. Lanham MD: Lexington Books. 2014.

Brodesco, A. 'Quantum Leaks. Uses of Scientific Theories in Television Series' *The Journal of Popular Television*. 2017.

Brown, P. 'Narrative: An Ontology, Epistemology and Methodology for Pro-environmental Psychological Research' *Energy Research and Social Science*, 31: 215–222. 2017.

Bruner, J. *Actual Minds, Possible Worlds*. Cambridge, MA: Harvard University Press. 1986.

Burroughs, B. 'Streaming Tactics' *Platform Journal of Media and Communications*, 8(1): 56–71. 2017.

Calik, S. and Corbacioglu, S. 'Role of Information in Collective Action in Dynamic Disaster Environments' *Disasters*, 34(1): 137–154. 2010.

Chayko, M. *Superconnected: The Internet, Digital Media and Techno-Social Life*. London and Thousand Oaks CA: Sage. 2017.

de Certeau, M. *The Practice of Everyday Life*. Berkeley, CA: University of California Press. 1984.

Goldsmith, J. and Wu, T. *Illusions of a Borderless World*. Oxford: Oxford University Press. 2005.

Gunduz, U. 'The effect of social media on identity construction' *Mediterranean Journal of Social Sciences*, 8(5): 85–92. 2017.

Harris, P. G. *What's Wrong with Climate Politics and How to Fix It*. Cambridge UK: Polity Press. 2013.

Kaapa, P. *Environmental Management of the Media: Policy, Industry, Practice*. London: Routledge. 2018.

Kim, E. 'In *Breaking Bad*, Why Is Walt Obsessed about the Fly in his Lab?' *The Blog* 28 December 2012 updated 27 February 2013. Available online at: www.huffingtonpost.com.

Laboto, R. and Thomas, J. *The Informal Media Economy*. Cambridge UK: Polity Press. 2012.

Mattenson, L. 'The Message of Breaking Bad: Live a Creative Life, not a Fantasy'. In Paul G. Harris, *What's Wrong with Climate Change and How to Fix It*. Cambridge UK: Polity Press. 2013.

Maxwell, R., Roundalen, J. and Vestberg, N. L. (eds) *Media and the Ecological Crisis*. London UK: Routledge. 2015.

McCloud, S. *Understanding Comics*. New York: Harper Collins. 1993.

McDonald, K. and Smith-Rowsey, D. (eds) *The Netflix Effect: Technology and Entertainment in 21st Century*. London: Bloomsbury. 2016.

Moezzi, M., Janda, K. J. and Rotmann, S. 'Using Stories, Narratives, and Storytelling in Energy and Climate Change Research' *Energy Research and Social Science*, 31: 1–10. 2017.

Morton, T. *Being Ecological.* Cambridge MA: MIT Press. 2018.

Murphy, P. *The Media Commons: Globalisation and Environmental Discourses*. Chicago: University of Illinois Press. 2017.

Murray, J. *Hamlet on the Holodeck.* Cambridge, MA: MIT Press. 1997.

Nelson, D.J., Grazier, K.R., Paglia, J. and Perkowitz, S. (eds) *Hollywood Chemistry. When Science Met Entertainment.* Washington DC: American Chemistry Society. 2013.

9 Video games and environmental learning

New modes of audience engagement

Overview

Much scholarship at the convergence of game studies and ecocritical inquiry investigates how digital games can be used to reinforce as well as reimagine existing environmental politics (Nguyen 2017). Digital games reached 74 billion players worldwide in 2011 and is expected to grow steadily into the future. There are over 500 million players of social games worldwide. The deeply engaging, interactive, spatial and temporal features of digital games have inspired marketers to attempt to develop equally immersive advertising content (Grimes and Shade 2005, cited in McAllister and West 2015). By 2006, for instance, 77 million has been spent globally on advertising in video games (Grimes in McAllister and West 2015: 386). One presumes this figure has dramatically increased over the intervening years. Consequently, it appears obvious that environmental media scholars need to examine how such a massive growth in this industry might communicate complex issues around climate change. What better format for illustrating the consequences of climate change or engaging with any other man-made disaster for that matter, than an immersive video game.

Yet, as John Parham notes, if environmentalists loathe the media, much of that loathing has latterly been aimed at computer games, habitually invoked as representing all that is wrong and all that is lost, in our relationship with nonhuman nature. Richard Louv's *Last Child in the Wood* in particular indicts computers, mobile phones and game consoles as part of his thesis that the generations of children born since the 1970s suffer from what he calls 'nature-deficit disorder' (Louv 2008). Meanwhile, at another extreme, Sean Cubitt has argued that technology is so intrinsically instrumental (2005: 4), if imbued with an ecological understanding, then mediated and technological culture almost counter-intuitively can be a place from which a renewed, revitalised, more complex sense of human ecology might emerge. This study seeks to call attention to the populist middle ground of media production and especially consumption, while concentrating on video games in this chapter. Parham usefully posits that such games can 'contribute both to a pragmatic understanding of and instruction in ecological issues, such as sustainable development or energy supply and to constituting or shaping environmental or ecological awareness' (Parham 2016: 205). This provocative approach will be followed through and illustrated by using examples that highlight areas of interaction, immersion and engagement that in turn might help kick-start a nascent environmental register for analysis and engagement, which the act of gaming can consistently offer to the growing numbers of players.

One of the first ecocritical essays to examine computer games was H. Lewis Ulman's 'Beyond Nature/Writing: Virtual Landscapes Online, in Print and in "Real Life"' (2001). This paper suggested that the 'virtual is always embodied in the real, just as the real is always mediated for us by the virtual' (348). This helps to support the proposition that *Pokémon Go* and other games like it, can usefully be examined as proto-typical environmental games. By all accounts, new modes of representation allow audiences to re-interrogate 'older representations of landscape' that through convention have become 'transparent' (345). Ulman posits computer games as 'mental models' or 'virtual models of one's environment' that promote forms of adaptation that 'tease out how we might co-exist with new ecological realities' (in Parham 2016: 208). Meanwhile, for other game scholars like Chang, games tackle ecological questions in a 'potentially less off-putting, less overtly didactic way' than other mainstream media, which hopefully will in turn 'encourage people to consider environmental problems and their solutions' (ibid.).[1]

Chang goes on to argue that computer games are different from other media because they move 'past the mere visualisation of data to procedural or algorithmic embodiment': specifically, the so-called realism of games resides not in graphics or narrative structure, but in what Ian Bogost calls 'procedural rhetoric'. Essentially, we learn ecological principles through playing a game and unlike conventional texts, games demand action. Games are 'richly designed problem spaces' or even 'possibility spaces' in which audiences and players come face to face up to their own knowledge of and impact on the environment (Chang 2011: 63, cited in Parham 2016: 208).

At the same time, one calls to mind Timothy Morton's provocative thesis that because of our idealisation of 'Nature', this has become a factor in what distances us from it. Through a process of 'othering', a 'really deep ecology' would 'jump down into the mud' (2007: 205). This dark ecological proposition – i.e. the sense of estrangement created by for instance the alien nature of othered animals and nature represented in games – can in turn lead to 'more stress, more disappointment, less gratification … and more bewilderment' (Morton 2010: 135). Consequently, Morton believes if we 'politicise' rather than 'aestheticize' or allow ourselves to wallow in, as he calls it this dark ecology. This approach will ultimately help us learn how to co-exist with the otherness of nature (see 2007: 113). While being a powerful and provocative argument, what is lost here according to Parham, is any sense that in structuring that political engagement, we will also need a visionary aesthetic or utopian imagination to go with this (Parham 2016: 205).

Philosopher Kate Soper (1995) has further conceptualised this proposition as the need to complement a 'nature scepticism', which actively interrogates the term 'nature' and the philosophies and practices we draw from it, with a 'nature endorsing' approach, which retains a sense of the power of nature and images of nature, 'to delight and instruct'. Or as other scholars suggest, while audiences need a critical perspective, they also need to retain a sense of the romance and excitement of nature ecology. In other words, scholarship needs to find ways to combine and crystalise the best attributes of first and second wave environmental criticism, which has remained a running theme through several chapters in this volume.

Offering another useful take, Christopher Hitt (1999) draws on the cultural conventions of the 'romantic sublime' which has been accused of generating images that implicitly assume mastery over nature, as evident for instance through the panoramic shots of many nature-focused features in Hollywood film (see Brereton 2005). Hitt's response to such images is more dialectical, suggesting the awe inspired by the sublime also tempers feelings of mastery or transcendence and indeed estrangement. By inspiring audiences and users, the sublime can offer kinship, even as we recognise that nature will always remain outside or 'other' to us as humans. Hitt quotes Scott Slovic in this regard: 'By confronting face-to-face, the separate realm of nature, by becoming aware of its otherness, the writer implicitly becomes more deeply aware of his or her own dimensions, limitations of form and understanding and processes of grappling with the unknown' (cited in Parham 228; Hitt 1999: 612). While some might say such assertions are further evidence of romantically naïve, first wave environmental criticism, nonetheless such expressivity remains evident, not just in eco-literature, but also in the most interactive forms of new media, including so-called serious video games.

One approach to help foster game literacy is to build educational modes of engagement, where students/gamers enjoy playing as well as developing their own re-purposed games. Studies by Buckingham and Sefton-Green among others, actively look at game literacy in the context of other media literacies that already exist (Squire 2008: 656).[2]

The consequences of active participation in gaming practices for developing new forms of literacies, include so many benefits that have been noted by a growing scholarship in the field:

- a producing as well as consuming information
- b the strategic use and critique of digital spaces that promote their particular values
- c the ability to produce meanings across multiple representational forms – from texts, graphs, charts and games modifications
- d texts that circulate within and across communities
- e trajectories for participation in social systems (including game journalism)
- f the repositioning of written texts from objects of authority to resources that are used in support of (digitally mediated) practice.

(Squire 2008: 653)

Overall, a strategy of co-opting a general history of Media Literacy and teasing out how trans-national research centres can be used to mark the evolving and necessary greening of the curriculum which helps support the expansion of new forms of environmental literacy is certainly called for. This is evident for instance in major educational hubs like the *Centre for Media Literacy* together with the Yale study of environmental literacy: www.yaleclimateconnections.org/topic/arts/. Such sites can provide a much-needed overview of this growing field. However, to put specific flesh on general media literacy protocols and their specific application for the development of environmental literacy, this games chapter will begin its case study overview of the subject with a short examination of the phenomenal growth of Twitch TV, followed by an analysis of Machinima and finally using a case study analysis of the game franchise *Pokémon Go* from an environmental perspective. But first it is important to tease out some of the unique features and affordances of video game play in general towards facilitating new forms of environmental learning.

New potential environmental affordances and multimodal game play

The term affordance is both contested and continuously debated within multimodal research and has particular emphasis and currency within social semiotic approaches to multimodality.[3] The concept originated in the work on cognitive perception by Gibson (1997) and was later taken up by Norman in relation to design (1988; 1990). Following Gibson, Van Leeuwen (1998) also conceptualises affordances to express 'meaning potential' and to refer to the material and the cultural aspects of modes of media production. Taking a slightly different emphasis however, modal affordance is applied to refer to what it is possible to express and is easily represented through the media. Kress usefully positions affordance as a complex concept, connected to both the material, cultural as well as the social historical use of a mode (each of which is intimately connected) (cited in Jewitt 2017: 24).

Applying such notions of affordances and multimodal modes to video game like *Pokémon Go*, where players are enticed to physically leave their indoors surroundings and engage with nature at its most primal and elemental level remains a useful example to focus on. This remains especially interesting from an environmental perspective, while exaggerating physical movement for effect. The effort and effect of venturing outdoors 'into nature' is recognised in mental health circles as broadly beneficial, as players within the game world ostensible find some digital Pokémon lurking around outside. Physical and especially mental well-being and health are constantly correlated with going into nature and actively communing with one's habitat.[4] At a more mainstream level of engagement Hollywood narratives provide vicarious, yet tangible, experiences of wild nature. Meanwhile, new media and video games in particular actively promote engaging and probably more immersive and interactive experiences – especially through the development of more extensive and serious gaming processes.

Consequently, from an environmental communications and green games perspective, there is nothing to lose by actively striving to re-purpose and re-apply conventional gaming protocols to facilitate and encourage all strands of media users towards developing and supporting increased levels of environmental literacy. Certainly, as shown through games research, this approach could directly apply to those players with, for example, profound disabilities, alongside those who for various reasons are not actively connected with nature or who don't necessarily display a surfeit of biophilia (or love of nature). Creatively and aesthetically, encouraging new-generational audiences to collect and maybe even learn, if only vicariously from (scientific) aspects of 'nature', is demonstrated later within an analysis of *Pokémon Go*. The game is certainly beneficial in itself and this mode of active engagement might further be considered useful in developing a basic level of environmental literacy, for instance, demonstrating how precious flora and fauna fit into a defined habitat or landscape and exploring how at a macro level at least 'everything is connected'.[5]

A long-standing tradition within environmental communication has been the broad strategy and objective of providing lay audiences with information-based appeals to help trigger pro-environmental concerns and sustainable behaviour (Jarreau *et al.* 2017: 145). Most certainly a base amount of knowledge about one's habitat and broader environmental issues may be an important prerequisite and first step to engineer sustained pro-environmental behaviour (146). One might even hope that such an (adapted) educational game, through using sophisticated interactive technology might appear more useful in promoting active, even critical engagement with aspects of the player's habitat. Hopefully, with some pre-designed e-learning steers, this can help infuse a sense of sustainable and environmental understanding and responsibility for real-world dynamics.[6] For instance using the popular new generational audiences strategy of a 'rewards system' or response, or simply providing a sensory 'kick-back' on the interface console – as evident in almost all First-Person Shooter games – such techniques are far from boring and never perceived as in any way preachy, according to players.

This technological hope remains, however, a big jump of the educational imagination in the struggle to discover more mainstream applications. But maybe with growing digital capability these various connections can be drawn together, as illustrated through a growing back-catalogue of serious educational game development. The ongoing concerns and moral panics about 'mindless fun' must be recognised, while also recognising the growth of material e-waste, which remains a key cautionary environmental concern for all media production and its consumption, if the medium is to carve out its authentic and hopeful critical green educational voice. If not with this piece of interactive software and new technology, maybe with some soon to be discovered virtual reality format or Killer App, this will become most effective. Concurrently, major strides can be made towards developing new forms of educational literacy and learning. A breakthrough could be made, for example, in the near future that specifically helps to marry advanced technology and promote robust environmental learning and active engagement while augmenting the importance of greening video games.

Video games and virtual reality: the intrinsic power of 'virtual nature' (album)

> To many who care about the natural world, no modern phenomenon seems more troubling than the emergence of 'virtual reality' as a new form of human experience. What only a few years ago seemed like science fiction now begins to seem an ever more plausible reality: the ability of people to experience 'environments' that are completely constructed by computers.[7]

Whether playing a computer simulation game, cruising the internet, or enjoying the ever more sophisticated fantasies that emerge from the studios of Industrial Light and Magic, adults and children alike are spending increasing amounts of time in cyberspace, isolated from real and first-hand exposure in nature. And yet it is also true that our awareness of potential environmental problems has an increasingly virtual quality, as we turn to computer models and simulated ecosystems to try to understand the complex changes going on around us in the natural world.[8] Drawing on this ever present boundary between 'real' and 'virtual', nature turns out to be rather more difficult, 'rather more revealing and instructive concerning our ideas of nature, than we might first have thought' (Cronon 1996: 439).

By all accounts, the digital games industry is part of the digital economy, but its boundaries are amorphous and poorly understood, according to Irish game scholar Aphra Kerr. The European Commission incidentally now refers to the sector as the Culture and Creative Industries (CCIs). While console games remain significant in mature markets, in newer markets mobile gaming on smartphones are demonstrating high levels of growth (Kerr 2017: 2). We are certainly seeing an increase in live performance of games through e-sports and the monetisation of play by players through broadcast channels, like gaming channels that are modelled on traditional type television channels, most notably Twitch, which is also archived on YouTube (3).

Industry analysts and professionals alike agree that the global market for video games will reach 100 billion dollars by the end of 2017 (NewZoo 2016) and that, thanks to localisation, 30–50% of that revenue is already coming from foreign markets (Chandler and Deming 2012).[9] Yet from the outside, and with my non-gaming hat on, it could appear that less risk is being taken in creating new and innovative games with fewer 'big games' like the highly innovative *The Last of Us* being developed. Similar to Hollywood at present, more of the percentage of total money spent is given over to supporting fewer bankable and broad-appeal global blockbusters, like the current proliferation of Marvel super-hero sequels on the big screen – all the while slavishly holding on to a predictable formula in gauging audiences' interests and pleasures.

Meanwhile, game scholars are concerned that part 'of the challenge is to maintain a level playing field for all players and in constantly tweaking game play and game mechanics to make it more interesting for returning players' (Storm 2012). Alternatively, academics like Ian Bogost (2010) have castigated the growth of social network games, affirming that in such games, not just nature but 'friends aren't really friends; they are just resources'.[10]

Of late, multinational global media technology companies like Apple have started to promote 'pay once and play' games in their app stores, returning to an older business model. In Europe, regulators have raised questions about free-to-play marketing and their monetisation techniques, especially when advertised specifically at children. Do 'freemium' or 'free-to-pay' games that include in-app purchases constitute false advertising one wonders, or more pointedly, when do social games and in-app purchases become a clear form of gambling (Kerr 2017: 113)?

By 2016, revenues from mobile games will surpass revenues from console games in many markets (NewZoo 2016, cited in Kerr 2017: 175).[11] The top freemium games are making significant revenues, with one report noting that *Clash of Clans* has 33,700 downloads and 168,000 dollars revenue per day in the US on the app store (Starkell 2014). While solo gaming remains an important part of this recreational experience, multiplayer matches and in-game chat prolongs the success of games, enlarging fandoms across countries and languages, which in turn highlights the social aspects of entertainment software. Furthermore, it has become clear that the largest and fastest growing markets are in Asia and especially in China, while the fastest growing mobile markets are situated in India, Indonesia and the Philippines (Kerr 2017: 176).

All the while, one could argue, cashing in on representations of nature and the environment remains big business. For example, recalling the treatment of nature as a conquerable resource in Sony Online Entertainment's fantasy-themed multiplayer role-playing game *EverQuest* (1999) (Stumpo), or in Zynga's farming simulation game *Farmville* (2009). Although such games can potentially serve to reinforce environmentally harmful attitudes, scholars like Parham have counter-argued that digital games can productively intervene in active ecological issues, 'including addressing renewable energy (Abraham), urban sustainability (Springer and Goggin), entanglement of nature and technology (Chang *et al.*), alongside human-animal relations (Attebery)' (Nguyen 2017: 22). At a stretch, however, digital games can further 'contribute both to a pragmatic understanding of and also providing instruction in ecological issues, such as sustainable development or energy supply and to constituting or shaping environmental or ecological awareness', all the while addressing these issues as 'material entities that need disposal and poses significant environmental problems' (Parham 2016: 205). Nevertheless, Parham remains fully aware of the paradox around the ever-present environmental problem of e-waste, noting the full life cycle for phone and computer devices that host so many games involves processes of extraction, production, consumption, obsolescence and disposal that enact harm to sites and subjects all over the globe (Nyugen 2017: 19).

Hitting the sweet spot: changing audience practices and environmental values

Such intractable political economy and resource waste issues need to be always kept in the forefront of critical analysis, but always focused from a player's perspective. In general, all types of games are designed to 'hit the sweet spot' between being too difficult or too easy, which in turn can also be applied to dealing with the complexities embedded within a broad range of environmental discourses.

More so than other new media formats, video games have become a form of interactive media which incorporate both closed and open-ended problem-solving tasks, requiring navigation across complex systems and encouraging various forms of productivity across digital media. Indeed, research comparing video-game playing practice to traditional academic literacy and technology standards, suggests that gaming is predominantly a literacy practice, requiring players to produce meaning from texts and produce 'expressive manifestations of technology across multiple forms' (Steinkuehler 2005, cited by Squire in Coiro *et al.* 640).

Meanwhile, according to many researchers the task for games scholars is not to argue specifically how games necessarily fit within and across traditional forms, but to better understand how games are converging with and transforming earlier forms of media expression. Examination of performance, interactivity and flow[12] among other attributes, characterise some of the key attributes of gameplay, which in turn can be scrutinised towards helping to tease out how these unique affordances might support an environmentally conceived register.

At their most elemental level, video games represent an interactive medium that draws most from the specific capabilities of the computer and, according to several scholars, in understanding them we must understand what it means to think, act, and learn in simulated worlds. However, a key question for this study remains focused around the implications of such an interactive medium for developing new forms of environmental literacy.

Ostensibly, we need not trouble ourselves to drive to the cinema to view a film any more. In a similar way, the avatar in a chat community is *e-moting* for us, externalising and embodying our very emotions. Video games offer numerous examples of such sensualised affects, but are often regarded as a post-emotional kind of experience. The game machine 'enjoys' our affective emotion through the avatar(s), who in turn represent us as players.[13] It is further suggested that the audience players simply invest their emotions in these cyber-agents. For instance, 'action video games bombard the console-player with their hyperaesthetic effects of over-presence' (see Jagadzinski 2004). But how can such unique affordances be further re-purposed to support an environmental and literary agenda? This remains an ongoing preoccupation of environmental and serious game studies. In trying to nail down specific affordances, it is helpful to highlight the exponential growth of the game platform, which coalesces across various levels of 'professional' game play, as a means for (amateur) audiences to enjoy and learn from other performances. To illustrate this phenomenon, lets now turn to a very successful format of game play afforded by Twitch.TV.

Twitch.TV and celebrity performers creating new forms of environmental learning

Jeff Bezos, the founder and CEO of Amazon.com, after buying Twitch.TV for over one billion dollars, has brought game play and its performative heuristics into the mainstream and communication scholars can certainly learn a lot from this growing phenomenon. 'Broadcasting and watching gameplay is a global phenomenon and Twitch.TV has built a platform that brings together tens of millions of people who watch billions of minutes of games each month' (cited in Forbes 2014). Launched in 2011, Twitch.TV was an experimental spin-off website, while by 2014 was using the eight most popular internet bandwidth out of all websites in the world (Keng 2014, cited in Anderson 2017: 1). A phenomenal success story and confirmation of expediential growth by any measure of comparison. In itself this justifies its focus for analysis, especially in illustrating its close engagement with new forms of game literacy.

Gaming has become a spectator sport, as envisaged by *Twitch* which is still a relatively new streaming video platform that broadcasts the best world players competing against each other. Many popular games, like *Call of Duty: Black Ops 2* (2012), offer a built-in theatre mode and webcasting function to allow players to capture and broadcast their gameplay through the company's servers. Hence the industry has quickly adapted to the notion of celebrity game players, recording themselves for mass observation and re-viewing. According to Hector Postigo (2014), many top YouTube commentators and performers can be characterised as 'hard-core gamers' who spend significant amounts of time playing and putting their money on peripherals to improve the entertainment value for their audiences (Kerr 2017: 134).

While Game Studies continues to focus on the corporeal element of games and identity formation, especially through gender and race (Shaw 2011); bringing all of this into the foreground with engagement through immersion and interactivity is another matter (Williams 2013). This overview chapter remains primarily interested, of course, in how an examination of such public performance of game play might inspire and signal new modalities of environmental connectivity or even call attention to new forms of active literacy.

Apparently, gamers are not satisfied with just the game, they are interested in the people surrounding the game world-view and the players at the helm, as well as the viewers who appreciate the performance. Certainly, one of the most compelling aspects of gaming involves how people play, view, comment and otherwise relate to each other.[14] Some further questions which can be used to make connections with and push developments in environmental media literacy, alongside new forms of audience interactivity could include:

- how do game streaming viewers think or feel about interactive online corporeality on game streaming websites?
- why do video streams of players in particular, hold an especially important place in game streaming, above other strategies aimed at bringing awareness to bodies?

- how are culturally coded aspects of corporeality, such as race and gender (and I would include a more broadly focused environmental predisposition), become re-constructed through game streaming?

(Anderson 2017: 13)

One wonders if these and other questions can be re-applied and re-formulated towards engaging with and promoting some form of prototypical environmental and representational literacy, all the while recognising new potential (green) affordances across the broad range of critical engagement with game play and most recently noting the enormous growth of celebrity game players.[15] But to answer these and other questions, much-needed empirical audience and behavioural research is required, which unfortunately is beyond the range and scope of this particular volume.

Meanwhile, at another end of the spectrum, Machinima takes active engagement in gaming to a whole new level. Both Twitch.TV and Machinima can be equally appropriated to encourage fruitful engagement with new forms of environmental learning. Cross media convergences enable platforms like Twitch.TV and new modalities of playing games like Machinima can be repurposed towards providing content for media online providers and in turn help develop a 'common culture' around a growing corpus of online and digital texts, while also developing best practice modalities for user engagement.

Machinima and new modes of engaging with environmental games

Since the rise of digital games, with their sometimes-complicated scenarios and narratives, coupled with their constantly extending online resources, while taking into account how new technologies are developed with their inherent affordances around interactivity, there has been growing interest in learning from such media how to produce interactive audio-visual content. Experiments applied have involved the use of databases containing pre-recorded, scripted video and audio files that would respond to the requests of viewers (Reinhard *et al.* 2017: 79). Witnessed at a very basic level in some cross-platform reality television series, for example, such experiments can be more fruitfully exemplified by the evolution of Machinima.

Similar to watching and performing games, as earlier illustrated by Twitch.TV, Machinima represents another important offshoot of video game fan culture. Through the manipulation of video game engines (the architectural code of a video game), players take control of the characters and use them to create short animated films within the game's 3D virtual environment – like, for example, in Aphmau's *Minecraft Diaries* running up to three series. Such experimentation recalls the creative and educational benefits of mash-ups of online adverts, like with Irish farming and an advertising campaign by 'Origin Green' discussed in Chapter 3. This use of new media affordances is pushing the capability of users to actively engage with such outputs. Creators of machinima films engage in transformative play – an act of altering the rules and structures of designated spaces to suit the player's individual needs. Unlike simply performing a game – no matter how many levels of complexity it entails – such proactive interventions add further layers of complexity, which can coincidentally be adapted to embrace the need for counter-hegemonic narratives that in turn speak to very radical environmental agendas. For instance, these could include topical environmental debates around 'the circular economy', 'de-growth' and other political and social strategies needed to support a move to a low-carbon transitional society. Most radically such engagement could assist in helping to de-centre the anthropocentric privileging of humans, as the only species worth protecting on the planet.

Because gamers are capable of fundamentally manipulating the medium of the video game in ways that other fans of traditional media cannot, notions of 'fan resistance' and creative re-imagining of habitats and human agency ought to be re-examined and teased out within this new context. Moreover, transformative play (a term borrowed from game design theory) most radically offers new ways of understanding consumer-producer relationships (Jones in Hellekson and Busse 2001: 261). All of which can restore innovative ways of perceiving and using game play, which with careful re-development can be further re-purposed to support new forms of environmental literacy.

This form of mastery over computers serves as a defining quality of hacker culture, which is a precursor to the contemporary fan culture of machinima. As an interactive medium, the video game requires the participation of the gamer. This disrupts the normal relationship between the media consumer and media producer. Jenkins' seminal study *Textual Poachers* (1992) refers to this phenomenon as 'participatory cultures' that engage players in textual poaching for their own purposes. Through the use of advanced skills, gamers can modify, or mod a game to suit their wishes and needs. For instance, at its most crass level, this might consist of removing the clothes of a character like Lara Croft. Alternatively, they can more usefully use these skills to manipulate the game engine and create radically progressive machinima relationships (Jones in Hellekson and Busse 2001: 264).[16]

Mojo.com talks, for example, of the most addictive games being *Civilisation*, which is packed to the brim with possibilities of creating a new/old world society and dramatising the eternal struggle to survive is evident in the cult global televisual franchise *Game of Thrones*, followed by the *CandyCrush* saga which was designed simply to 'kill time' as an easy-to-use free game on the mobile phone. Of late, it has become globally addictive and players can end up buying a lot of additional features to improve its playability. Meanwhile, apparently, the top two addictive-games are *Minecraft* which is a sand-box game, where you build your own world, and *World of Warcraft* that has been enticing fans for over a decade now and provides so much complex engagement with ongoing action and adventure.[17] Most of these games, however, simply use nature as a resource to be exploited, while such a regressive evocation of environmental usage can at the very least bring these issues to the attention of users.

Consequently, if we understand the video game not merely as a medium of (passive) consumption, but also as a means of prosumer production, the communities it creates through various forms of mods can seem proactive as a consequence and it is certainly useful in promoting various forms of critical literacy. Manovich (2001) most successfully suggests that the process of navigating space is a defining characteristic of new media and does not adhere to the viewing expectations of the watchers. This is especially apposite with the proliferation of first-person shooter games. Galloway (2006), for instance, traces the history of the first-person perspective in film with the 'subjective shot', citing early filmic examples like *The Lady in the Lake* (1946) and *Dark Passage* (1947). He goes on to say:

> more often than not, this type of shot is used to show the vision of the criminals, monsters or killer machines. This useful analysis shows that the merging of camera and character in the subjective shot is more successful if the character in question is marked as computerized in some way.
>
> (12)

One wonders if such first-person shooter type aesthetics could for instance help promote and extend environmental literacy (Jones in Hellekson and Busse 2001: 270).[18] For instance, *The Long Dark* (above) and *Rust* (2014) are both first-person shooters in which players survive in hostile environments. While *The Long Dark* is a single player experience and *Rust* is multiplayer only,

> the only aim in *Rust* is to survive. To do this the player will need to overcome struggles such as hunger, thirst and cold. Build a fire. Build a shelter. Kill animals for meat. Protect themselves from other players and kill them for meat. Create alliances with other players and form a town. Do whatever it takes to survive.'[19]

All of which actions constitute a survival response to living in a hostile environment, where players have to learn how to cope with varying degrees of difficulty. At the same time these actions in coping with a hostile environment also serve to teach a number of cautionary environmental agendas. The overall diegesis of the game remains, however, totally constrained and somewhat arbitrary. Meanwhile a paradigm shift I suggest has been created with *Pokémon Go*, which we now turn to. Its direct real time access to, and usage of, one's real life habitat means it becomes directly part of the diegesis of the game play.

Pokémon Go: collecting nature, promoting environmental affordances

Calling attention to a recent very popular form of geo-location gaming mixed with AR, *Pokémon Go* can possibly with some imagination be re-purposed for a new generation to help connect with and speak to a form of biophilic 'nature', while also calling attention to a growing health worry by scholars and parents around new generations spending too much time indoors on their digital media games consoles.

Pokémon Go essentially constitutes an augmented reality and hybrid form of (AR) which bring the real and virtual world together and can potentially at the same time make the real world appear less real. But then of course the opposite could also be the case. For example, with their smartphone game-aesthetics, such games replicate many aspects of real-world wildlife watching and natural history, by allowing players to find, capture, and collect Pokémon, who are effectively virtual animals. Using the smartphone's GPS system together with Google Maps, the game provides users with an augmented reality experience where they encounter, catch and collect virtual species of Pokémon, while coincidentally if vicariously exploring their own real world (Dorward *et al.* 2017: 2).

As Satoshi Tajiri noticed during the 1990s, as rapidly growing urban areas offer limited opportunities to connect directly with nature (Allison 2003), augmented digital worlds might in some way compensate for their obvious limitations. In essence, more positively environmental communications scholars might suggest that the game could be repurposed in some way to help its users reconnect with nature and help stave off any sense of nature deficit disorder. This emanates from a widespread concern among conservationists and environmental activists alike, as discussed earlier, that people, and particularly young people, have become disconnected from nature through urban living and are therefore less likely to value wildlife and wild places, as suggested by MacFarlane to be explored later (Balmford *et al.* 2002).

A tangential connection to this particular game was sparked by a 2017 trip to Shrewsbury in the UK, where Charles Darwin hailed from. The local museum has an exhibition of a number of bird's eggs, plants and stuffed animals collected from the period. As Susan Sontag has most famously argued, photography helped replace such invasive, if direct engagement with nature through pastimes like hunting and shooting. This sparked my imagination, wondering if this new global phenomenon of *Pokémon Go*, involving physically going out into the neighbourhood landscape with one's mobile devise to find some illusory Pokémon, has gone a step further.[20] With some innovative dexterity, might the gaming phenomenon be further re-imagined and re-booted as an educational environmental game-structure, which could usefully provoke a deep engagement with raw nature? A big ask I know! Possibly, my game experts tell me, by using image recognition software to 'catch' actual animals, this educational strategy could be re-activated and realised in the near future.

This is evident, for instance, in other games like *The Last of Us* or *Zombies, Run!* but without the obligatory fear and horror around such 'othering' of species in this audio driven running game. 'Get ready for the run of your life. Join two million runners on an epic adventure! You tie your shoes, put on your headphones, take your first steps outside. You've barely covered 100 yards when you hear them. They must be close. You can hear every guttural breath, every rattling groan – they're everywhere. Zombies. There's only one thing you can do: Run!'[21]

Far from the phenomenal rise of zombie culture, concern has been constantly expressed that interest in natural history is fading (Tewksbury *et al.* 2014) and that skilled natural historians and even taxonomists are in increasingly short supply. In any case the (post-colonial coded) game encourages people to get outside into the real-life world-view of nature. However, by promoting the idea of 'catching' creatures that are subsequently used to fight against each other, the game may at the same time help reinforce utilitarian and exploitative relations between human and nonhuman nature. This is evident in earlier historical exhibits in museums, while contemporary digital games afford examples of real animals being 'caught' in 'Pokeballs' on social media.[22]

Sadly, the brightly coloured, exciting and easily accessible Pokémon species may alternatively distract people from recognising real species and the problems they face (Sandbrook *et al.* 2015). Furthermore, it has been argued by some studies that conservation efforts to reconnect people with nature can have the opposite of what might be intended, because constructing the problem as one of dissociation with nature reinforces the idea that humans occupy a distinct category from nonhuman nature (Fletcher 2016, cited in Dorword *et al.* 2017: 5).[23] These are all valid concerns, but again these worries can only be fully measured and teased out through extensive audience studies to help determine a particular game's usefulness in promoting a critical engagement with nature.

Meanwhile, at a populist level of engagement, most media and journalistic debate around the summer phenomenon recalled the off-repeated moral panic and fear around sedentary and vicarious engagement. Such debates are constantly signalled by the (anti-environmentalist) pleasures of young people being hypnotised by artificial video-game play, which is usually carried out in the safety of indoors environments. Mark Balnaives *et al.* (259) invoked *Pokémon* 'as an example of the kind of transformational multimedia phenomenon that students of media, culture, and society will be increasingly faced with in the 21st century' (see Peterson 2003: 14). All the while, the unique game activity in *Pokémon Go* appears to subvert the normal stereotypes of new digital technology and its otherwise artificial affordances, while at the same time maintained its core attributes of immersive gameplay.

Since *Pokémon* was originally released in 1996 for the Nintendo Game Boy, the links between games and collectables has remained strong. Sara Grimes (2015) has detailed how this connection has operated in games aimed at children, including *Webbinz, Club Penguin, Moshi Monsters*. In a very interesting piece by Robert MacFarlane for BBC's *Newsnight* (4 December 2017), the author spoke of the poetry book he wrote entitled *The Lost Words*, which was illustrated by the artist Jackie Morris.[24] The project was inspired by his shock at a paper in the academic journal *Nature*, which suggested that the Pokémon phenomenon meant that children were losing their connection with nature, which remains a constant refrain across popular discourse. The academic study found that children spend too little time 'in nature' and could recall only the names of Pokémon characters, instead of real-life nature plants and animals. MacFarlane's counter-hypothesis suggested that 'good names help us see and help us love nature', echoing Louv's 'Nature Deficit Disorder' thesis. He concluded with a memorable assertion 'what we do not love, we do not save' (as discussed on *Newsnight*, 4 December 2017). This corresponds with a long help belief, embedded within first wave environmental discussions around loving nature, (as discussed in Chapter 3 and elsewhere) as being a central prerequisite for developments across new forms of active environmental literacy

The artist alongside the environmentalist bemoans the fact that publics (especially young people) are forgetting the names of real plants and animals. Alternatively, when children are encouraged to venture into nature, such as a forest, they become prototypical 'naturalists' and can soak up the 'meaning of nature'. Recounting a dwindling of natural species through biodiversity reduction, MacFarlane believes *we* are most certainly adding to this loss, by not being even able to name what is being lost – recounting for instance 'the difference between a starling or a blackbird or know what an ash tree looks like' (also from the discussion on *Newsnight*, 4 December 2017). Names, alongside visual representation and allegorical metaphors through a range of audio-visual media, certainly mean a lot and help in supporting the basic heuristics around environmental education.[25]

Concluding remarks

Louv's 'natural deficit disorder' has increasingly afflicting the American (and global) population, according to scholars like Robert Fletcher who goes on to claim this condition is attributed to growing urbanisation, to an increase in 'videophilia' (reliance on electronic media for virtual nature experiences) and to decreasing opportunities for experiential environmental education in school curricula (Kareiva 2008). But contrary to this type of logic and environmental reasoning, the analysis developed by Fletcher and others support Russell's (1999) contention that the 'proper' concern for nonhuman nature that would end up inspiring the type of environmentalism that in turn support a desire to visit national parks and other protected areas, may be a function of a particular cultural perspective, rather than an objective relationship with 'nature' per se. In other words, for contact with nonhuman nature to have the desired effect of inspiring support for environmentalism, it may necessitate the capacity to see and know nature in a particular way (Argyrou 2005). Hence, when Kareiva (2008) laments that 'humans are becoming seriously disconnected from nature' (2757), he may be in essence expressing a class-based culturally specific viewpoint concerning humans' relationship with the rest of the world, rather than pronouncing on an objective, universal condition (Fletcher 2015: 345). Nevertheless, one could postulate that serious learning games – which in turn privilege nature and science – can at least actively help build the potential for solid understanding and appreciation of our natural world.

By re-valuing the benefits of synthetic media and video games, as a prototypical mode of learning environmental literacy, this can be argued as an important first step. Fletcher further recalls the commonly held presumption that is 'increasingly common among conservationist and environmental educators, that more direct nature experience will inspire pro-environmental behaviour.' While agreeing that 'the interrelationship between "experience" and prior "stories" that participants in environmental education activities carry with them, helps to determine how they are affected by their participation', nonetheless 'substantial conceptual work is necessary in order to invoke the proper experience inspiring the type of action environmentalists desire' (Fletcher 2015: 346).

This book, and this chapter in particular, might hope that the artificial or synthetic nature of audio-visual media and their powerful gaming story-worlds are at least potentially capable of affording 'proper experiences' and re-constituting nature most effectively for such an educational agenda as Fletcher suggests. Environmental scholars rightly question easy binary oppositions between digital nature and so-called first-hand venturing out into nature – recalling for instance the famous philosophical allegory of Plato's Cave. Yet one could infer that the reflection on the cave wall can equally serve as a potent signifier, if audiences are especially in the right frame of mind, to decode such imagery through an environmental lens. The power of narrativising nature within games in particular, can help provoke an almost epiphytic engagement with or connectivity towards natural organisms and issues that hopefully in turn can be highlighted, sometimes more effectively at a distance through an intermediary digital medium. This hope is illustrated through successful games like *Pokémon Go* as well as others like *Farmland.*

Notes

1 See 'Playing the Environment: Games as Virtual Ecologies' in *Proceedings of the Digital Arts and Culture Conference 2009 – After Media: Embodiment and Context* (Parham 2016: 208).

2 Different ways of experiencing the same fictional world, including for example, *Game of Thrones*, recalling the intertextuality of the novels, TV series, games, etc. or *The Walking Dead*, which began as a comic, followed by TV series, and other spinoffs and a video game. This continuity and cross-fertilisation between a range of literacies also remain a working hypothesis of this study.

3 A pervasive criticism of multimodality is that it can seem rather impressionistic in its analysis. How do you know that this gesture means this, or that this image means that? In part, this is an ongoing preoccupation, recalling its linguistic heritage, while reflecting the proposition that the task of building 'stable analytical inventories' of multimodal semiotic resources is always up for contestation (Jewitt 2009: 26).

4 For instance literary and serious games analysis by scholars like Daniele Barrios-O'Neill who promotes very incisive cross-disciplinary engagement, which in turn has application for environmental literacy – as explored in Barrios and Hook 2016. While, for instance, typical interactive games may be comprised of design processes aimed at influencing players by stimulating introspection and self-reflection, 'serious games are designed to educate and influence players attitudes and actions in the real world, and are likewise discusses in terms of their effects on the ecological, cultural and financial systems that they simulate or critique' (125).

5 The latest 2017 *Star Wars* saga and its ur-mythic recuperative landscape of Skellig Michel in Ireland comes to mind, reinforcing and repurposing the tag line 'let the force be with you' to also address if not secure deep environmental connectivity.

6 See for instance 'The Gamification of Nature' which links to https://agentsofdiscovery.com/ described as 'An educational technology platform that empowers educators to create fun and engaging augmented reality games'.
7 Some of these environments come complete with their own cash economies – *Eve Online, World of Warcraft* (complete with real world illegal gold mining sweat shops in the developing world), *Roblox, Minecraft*, and many more. The Entropic Universe even has a real cash economy – see www.entropiauniverse.com/.
8 For instance, digital twin technology is an emerging technology in this area.

> Digital twin refers to a digital replica of physical assets, processes and systems that can be used for various purposes. The digital representation provides both the elements and the dynamics of how an Internet of Things device operates and lives throughout its life cycle.
>
> (See https://en.wikipedia.org/wiki/Digital_twin)

> The Internet of Things (LoT) involves the embedding of network technologies integrally into lived environments that allows everyday objects to become augmented with communications capacities and represents an interconnected, intelligent, real environment. All the while, video games are extremely well suited to this paradigm, being able to mount augments and enact realities in ways that emphasise our participatory roles in complex systems, while simultaneously interrogating the nature and effectiveness of those systems.
>
> (Barrios-O'Neill and Hook 2016: 127)

9 The top 10 countries with regards to game revenues (NewZoo 2016): America 20 m, China 18 m, Japan 12 m, Germany 3 m, UK 3 m, Republic of Korea 3 m, France 2.6 m, Canada 1.7 m, Italy 1.5 m, Spain 1.4 m.
10 See, for instance, *League of Legends: Summoner's Rift*, Riot 2009–2016.
11 Total consumer spends on games in America reached 22.41 billion in 2014 with four out of every five American households owning a device used to play video games. Digital natives have taken to such interactive media in high numbers. It is estimated that 59% of American citizens (ESA 2014) and 48% of European citizens (ISFE 2012) play video games regularly.
12 Csikszentmihalyi describes the phenomenon of flow – the condition under which a player's skill level is optimally pitched with the level of challenge faced, creating conditions in which the participant reports a feeling of being in the zone (Squire 2008: 643).
13 In *The Ambiguity of Play*, Sutton-Smith (1997) cogently argued for the centrality of play in human experience, contending that play has been conceived primarily along seven or so rhetorical lines: including the ancient four, (a) fate, (b) power, (c) community identity and (d) frivolity, and the modern discourses of, (e) progress, (f) imaginary, and (g) the self. Video games can bring all of these aspects to a whole new level.
14 This is pushed to another level for instance through the benefits of add-ons or bonus features for enjoying the pleasures of film (see Brereton 2012), where audiences can spend hours watching media creatives talk of how they made film and affording pleasure to their intended audience.
15 For instance, one of the most successful European live vloggers is Felix Kjellberg, who is known online as PewDiePie or 'pewds'. Psychology researchers in particular have shown that media messaging by such celebrity can help highlight pro-environmental social norms and significantly promotes positive behaviour towards for instance energy use (Goldstein, Cialdini and Griskevicius 2008, cited in Jarreau *et al.* 2017: 147). Furthermore, as reiterated across earlier chapters, environmental psychologists often indicate that talking about local issues can in turn trigger connections of place attachment and place identity. By using narrative hooks or cogent metaphors in

gameplay, they can certainly help grab audiences' attention by communicating environmental issues to which they can subsequently relate to and develop further (Jarreau *et al.* 2016: 155). Consequently, their celebrity status can garnish so much attention, no matter what games they play, while in turn helping to promote best practice and a sense of successful active engagement.

16 See, for instance, Zynga's farming simulation game *Farmville* (2009) which could most fruitfully be re-purposed to speak to environmentally sensitive food security debates.

17 In December 1993, id Software released DOOM which changed the face of computer gaming, games could both *play* the game and *play* with the game's design. Mods reveal one of the unique qualities of the video game as a medium – one that separates it from traditional forms of audio/visual media (Jones in Hellekson and Busse 2001: 266).

18 While film spin-offs from games have been surprisingly unsuccessful, maybe the *Angry Birds* movie might help break the mould. An analysis by Randy Nichols (2014: 121) found that the *Tomb Raider* film released in 1996 with Angela Jolie in the lead role has been the most successful of these film tie-ins to date, with over 100 million grossed at the box office. The Disney studio, in owning a long list of game studios and having extensive links between Pixar and game publishers in films like *The Incredibles* (2004), *The SpongeBob SquarePants Movie* (2004), *Spider Man 2* (2004) and *Wreck-it Ralph* (2012), most obviously highlights the potential synergies between film, animation and games (Kerr 2017: 115). The new *Tomb Raider* (2018) film, based on the recent *Tomb Raider* game reboot looks certain to top the success of this one. Available online at: www.imdb.com/title/tt1365519/?ref_=nv_sr_1.

19 Available online at: http://store.steampowered.com/app/252490/Rust/.

20 Danielle Barrios-O'Neill in private correspondence recalls the games possible link with citizen science – 'it kind of borrows the method of apps. that have people tag wildlife and at the same time there's something mildly colonial about it, in the way it enacts a collecting of species/resources from a natural environment. Depending on which way you choose to interpret the game mechanics, there are potentially two really different worldviews in tension there.'

21 Available online at: www.theverge.com/2016/11/30/13799582/amazon-rekognition-machine-learning-image-processing.

22 Available online at: www.google.com/search?q=Desejos+De+Homem+2016+real+animals+caught+in+pokeballs+on+social+media&tbm=isch&tbo=u&source=univ&sa=X&ved=0ahUKEwiw56Gr1qXcAhVEa1AKHQeTBzkQsAQIKQ&biw=1440&bih=691.

23 Some sites, building on the popularity of the 'Pokeblitz' concept (e.g. newly developed websites such as Pokemapper (www.pokemapper.co) and Poke Radar (www.pokeradar.io)) already map the 'distributions' of different Pokémon in ways that are striking similar to citizen science projects such as eBird (www.ebird.org) and iNaturalist (www.inaturalist.org).

24 Review in the *Guardian.* Available online at: www.theguardian.com/books/2017/oct/02/the-lost-words-robert-macfarlane-jackie-morris-review.

25 Pergmams and Zaradic (2008), for example, have researched the decline of visitations to American National Parks since 1987 and detected a significant relationship between this decline and 'videophilia', or the replacement of outdoor activities with endless hours spent playing video games and plugged into the internet. City-dwellers depend on nature for clean water, food, and climate regulation, but it is easy to lose sight of that dependence unless concerted efforts are made to educate urban-dwellers about the services that ecosystems quietly provide. A poor understanding of basic natural history is sure to undermine audience's ability to solve environmental problems. In a study of British school children between the ages of four to 11, they were twice as good at identifying characters from Pokémon than common organisms such as a beetle or a rabbit. People care about what they know, and people need to know something about nature to help solve environmental problems.

References

Allison, A. 'Portable Monsters and Commodity Cuteness: *Pokémon* as Japan's New Global Power' *Postcolonial Studies*, 6: 381–395. 2003.

Anderson, Sky La Rell 'Watching People is Not a Game: Interactive Online Corporeality, Twitch.TV and Videogame Streams' *Game Studies*, 17(1). 2017. Available online at: http://gamestudies.org.

Argyrou, V. *The Logic of Environmentalism: Anthropology, Ecology and Post-coloniality*. New York: Berghahn. 2005.

Backe, H.-J. 'Within the Mainstream: An Ecocritical Framework for Digital Games History'. *Ecozon@2017*, 8(2): 18–38.

Balmford, A., Clegg, L., Coulson, T., Taylor, J. and Street, D. 'Why Conservationists Should Heed *Pokémon*' *Science*, 295(5564): 2367. 2002.

Balmford, A., Bruner, A., Cooper, P. and Costanza, R. 'Economic Reasons for Conserving Wild Nature' *Science Compass Review*. 2002. Available online at: http://science.sciencemag.org/content/sci/297/5583/950.full.pdf?casa_token=_vkoh0IiAbAAAAAA:emxeZYpVryEI8SiBFZcts-sHKi5fJip8kLiu6m2dAQSkU13Un7GvPoMlocPfl0UDPs0Jv33FM.

Balnaives, M., Hemelryk Donald, S. and Shoesmith, B. *Media Theories and Approaches: A Global Perspective.* Basingstoke UK: Palgrave. 2008.

Barrios-O'Neill, D. and Hook, A. 'Future Energy Networks and the Role of Interactive Gaming as Simulation' *Futures*, 81: 119–129. 2016.

Bogost, I. *Persuasive Games: The Expressive Power of Video.* Cambridge MA: MIT Press. 2010.

Boyle, E, Connolly, T. M. and Hainey, T. 'The Role of Psychology in Understanding the Impact of Computer Games' *Entertainment Computing*, 2(2): 69–74. 2011.

Brereton, P. *Smart Cinema: DVD Add-ons and New Media Pleasures*. Basingstoke UK: Palgrave. 2012.

Chandler, H. M. and Deming, S. O. M. *The Game Localisation Handbook.* Burlington MA: Jones and Bartlett Learning. 2012.

Cronon, W. (ed.) *Uncommon Ground: Rethinking the Human Place in Nature*. New York: Norton and Co. 1996.

Cubitt, S. *Eco Media.* Amsterdam: Rodopi. 2005.

Dorward, L., Mittermeier, J. C., Sandbrook, C. and Spooner, F. 'Pokémon Go: Benefits, Costs, and Lessons for the Conservation Movement' *TOC,* 10(1): 160–165. 2017.

Esser, A., Smith, I. R. and Bernal-Merino, M. A. *Media Across Borders – Video Games.* London: Routledge. 2015.

Fletcher, R. 'Nature is a Nice Place to Save but I Wouldn't Want to Live There: Environmental Education and the Ecotourist Gaze' *Environmental Education Research,* 21(3): 338–350. 2015.

Forbes, T. 'Google Flinches: Amazon Grabs Twitch's 55 Million Monthly Viewers for 1B' *Marketing Daily*. 26 August 2014. Available online at: www.mediapost.com/publications/article/232785/google-flinches-amazon-grabs-twitchs-55-million.html?edition=75636.

Galloway, A. (ed.) *Gaming: Essays on Algorithmic Culture*. Mineapolis MI: Minnesota Press. 2006.

Gibson, J. J. *An Ecological Approach to Visual Perception.* Boston: Houghton Mifflin. 1997.

Grimes, S. 'Online Games and Children'. In Mansell, R. and Ang, P. H. (eds) *The International Encyclopedia of Digital Communications and Society*. Oxford: Wiley Blackwell. 2015.

Hannigan, J. *Environmental Sociology* Second Edition. London: Routledge. 2006.

Hellekson, K. and Busse, K. (eds) *Fan Fiction and Fan Communities in the Age of the Internet.* Jefferson NC: Macfarland Press. 2001.

Hitt C. 'Towards an Ecological Sublime' *New Literary History*, 30(3): 603–23. 1999.

Jagadzinksi, J. *Youth Fantasies.* Basingstoke UK: Palgrave. 2004.

Jarreau, P. D., Brown, P., Altinay, Z. and Reynolds, A. 'Best Practices in Environmental Communication: A Case Study of Louisiana's Coastal Crisis' *Environmental Communication.* 2017.

Jenkins, H., Ito, M. and boyd, d. *Participatory Culture in a Networked Era: A Conversation on Youth, Learning, Commerce and Politics*. Cambridge UK: Polity Press. 2016.

Jewitt, C. (ed.) *The Routledge Handbook of Multimodal Analysis.* London: Routledge. 2009.

Jones, R. 'From Shooting Monsters to Shooting Movies: Machinima and the Transformative Play of Video Game Fan Culture', 261–280. McFarlane and Co., 2006.

Juul, J. *Casual Revolution: Reinventing Video Games and Their Players.* Cambridge MA: MIT Press. 2009.

Kareiva, P. 'Ominous Trends in Nature Recreation' *PNAS*, 105(8): 2757–2758. 2008. Available online at: www.pnas.org.

Kerr, A. *Global Games: Production, Circulation and Policy in the Networked Era.* London: Routledge. 2017.

Louv, R. *Last Child in the Woods: Saving our Children from Nature-Deficit Disorder*. Chapel Hill, NC: Algonquin Books of Chapel Hill. 2008.

Manovich, L. *The Language of New Media*. Cambridge MA: MIT Press. 2001.

McAllister, M. and West, E. (eds) *The Routledge Companion to Advertising and Promotional Culture*. London: Routledge. 2013/15.

Morton, T. *Ecology Without Nature: Rethinking Environmental Aesthetics*. Cambridge MA: Harvard University Press. 2007.

Morton, T. *The Ecological Thought.* Cambridge MA: Harvard University Press. 2010.

NewZoo 'Top 100 Countries by Revenue'. Available online at: www.Newzoo.com (March 2016).

Nguyen, J. 'Digital Games about the Materiality of Digital Games'. *Ecoozon@,* 8(2): 22–38. 2017.

Nichols, R. *The Video Game Business*. London: BFI and Palgrave Macmillan. 2014.

Norman, D. *Affordances, Convention and Design*. New York: Basic Books. 1990.

Parham, J. *Green Media and Popular Culture*. Basingstoke UK: Palgrave. 2016.

Parham, J. and Chang, A. (eds) 'Green Computers and Video Games: An Introduction'. *Ecozon@*, 8(2): 1–17. 2017.

Pergams, O. R. and Zaradic, P. A. 'Evidenced for a Fundamental and Pervasive Shift away from Nature Based Recreation'. *Proceedings of the National Academy of Sciences of the United States of America* (PNAS), 105(7): 2295–2300. 2008.

Peterson, M. A. *Anthropology & Mass Communication: Media and Myth in New Millennium.* New York: Berghahn Books. 2003.

Postigo, H. 'The Socio-technical Architecture of Digital Converting Play into YouTube Money' *New Media and Society*. 2014. Doi: 10.1177/1461444814541527.

Reinhard, C.-L. and Amsterdam, P. 'A Community of Televised Avatars: Interactivities in Virtual World Television Promoting and Acknowledging Participatory Communities' *Participations,* 14(1): 67–87. 2017.

Russell, C. L. 'Problematising Nature, Experiences in Environmental Education. The Interrelationships of Experience and Story' *Journal of Experiential Education*, 22(3): 123–128. 1999.

Sandbrook, C., Adams, W. M. and Monteferei, B., 'Digital Games and Biodiversity Conservation'. In *Conservation Letters*. Wiley. 2015. Available online at: https://onlinelibrary.wiley.com/doi/pdf/10.1111/conl.12113.

Shaw, A. 'Do You Identify as a Gamer? Gender, Race, Sexuality and Game Identity'. *New Media and Society*, 14(1): 28–44. 2011.

Shephard, K., Harrraway, J., Lovelock, B., Skeaff, S., Slooten, L. and Strack, M. 'Is the Environmental Literacy of University students measurable?' *Environmental Education Research*, 20(4): 476–495. 2014.

Sontag, S. *On Photography*. London: Penguin. 1977.

Soper, K. *What is Nature? Culture, Politics and the Non-Human*. Cambridge UK: Wiley Blackwell. 1995.

Squire, K. D. 'Video Games Literacy: A Literacy of Experience'. In J. Coiro, M. Knobel, C. Lankshear and D. J. Leu (eds) *Handbook of Research on New Media Literacies*, pp. 635–669. London: Routledge. 2008.

Starkell, N. 'How Much Money Top 50 Free Mobile Apps Actually Make'. *Huffingtonpost.com*. 2014. Available online at: www.huffingtonpost.com/natasha-starkell/how-much-money-top-50-fre_b_5279520.html.

Steinkuehler, C. 'The Literacy Practices of Massively Multiplayer Online Games versus Online Standards'. Paper presented at the Annual Meeting of the American Educational Research Association, Montreal, Canada. 2005.

Storm, D. 'For Riot Games, Big Data is Serious Business'. *Insights.dice.com*. 2016. Available online at: http://insights.dice.com/2012/12/07/for-riot-games-big-data-is-serious-business/ 2012.

Sutton-Smith, B. *The Ambiguity of Play*. Cambridge MA: Harvard University Press. 1997.

Tewksbury, J. J., Anderson, J. G. T., Bakker, J. D., *et al.* 'Natural History's Place in Science and Society'. *Bioscience*, 64: 300–310. 2014. Available online at: https://academic.oup.com/bioscience/article/64/4/300/2754159.

Thevenin, B. 'Twine as Alternative Media: Video Games, The Culture Industry and Social Change' *Teaching Media Quarterly*, 5(2). 2017. Available online at: file:///Users/pat/Downloads/652-Lesson%20Plan-821-1-10-20171106%20(1).pdf.

Ulman, H. L. 'Beyond Nature/writing: Virtual Landscapes Online in "Real Lives"'. In Karla Armbruster and Kathleen R. Wallace (eds) *Beyond Nature Writing: Expanding the Boundaries of Ecocriticism*, pp. 341–356. Charlottesville: University Press of Virginia. 2001.

Van Leeuwen, T. 'It Was Just Like Magic: A Multimodal Analysis of Children's Writing' *Linguistics and Education*, 10(3): 273–305. 1998.

Williams, K. D. 'The Effects of Video Game Controls on Hostility, Identification, and Presence'. *Mass Communication and Society*, 16: 26–48. 2013.

10 Going viral

YouTube and new forms of environmental literacy

Overview: the online power of storytelling and narrative transportation

Gretter *et al.* (2017) convincingly argue that an essential step in teaching students to critically assess online content is to first understand how the human mind distinguishes between reality and fiction, when being transported through online storytelling spaces. Most specifically for this study, we must ask: 'How do digital stories impact on youth's sense of reality in speaking to environmental concerns'? This overview survey advances a step in this direction by looking at online stories through the lens of narrative transportation theory. Most specifically, Gretter *et al.* alludes to basic narrative rules of engagement: (1) what makes stories a powerful tool to convey ideas, (2) how digital storytelling differs from traditional storytelling and (3) what educators can do to help students critically assess stories in online spaces (2017: 2).

Storytelling in the twenty-first century has evolved as people have found new ways to record, share and consume stories: games, personal experiences, or news, being just a few of the examples highlighted in the literature (Lundby 2008). Today, modern technology has merged the visual, auditory and the textual, according to numerous observers, while giving everyone the potentiality to become their own storyteller and make use of their digital environment for personal purposes. Recent statistics show that more than 94% of tweens and teens use digital media on a daily basis, including watching online videos, reading, gaming, and interacting on social media, while 34% of tweens and 28% of teens write stories, articles or blogs themselves (Rideout 2015).

Millard (2005) emphasises how children's (and young people's) understanding of 'narrative forms in different modalities may support one another' (162). While Reid (2003) argues that film (which can also be considered as short-hand for all new media narrative formats) can be used to help scaffold effective writing and critical engagement and that students can learn about narrative by 'shuttling' between the two forms of film and print in order to engage with the different modes and affordances of each form. Reid usefully proposes that print and film studied together help to make explicit what they have in common and what is specific to each form. Such erstwhile old school 'adaptation' and multi-modal linkages, as well as transmedia protocols, can of course be extended to all new media formats.

Green *et al.* (2012) reminds us that psychological immersion into someone else's story – also called 'narrative transportation' – entails emotionality and highlighting an attentional focus, which becomes a very powerful tool in communicating with audiences. Two of their research studies suggested that when readers' pre-reading emotional states matched the emotional tone of a narrative, transportation into that narrative was increased (Gretter *et al.* 2017: 3). Of course, it makes sense that the mediated message has to both cognitively and emotionally speak to its audience to become fully effective. Consequently, examining the most popular or most innovative YouTube videos will help signal their unique aesthetic powers in addressing environmental concerns.

Many of the social networks and archival digital fora which are used to disseminate videos, computer games, music clips, memes or photos (Bott 2009), often rely more on visuals than on text-based narratives to relay emotive media messages, recalling the strategies of mainstream and niche media in spreading viral videos (Zelin 2015). Furthermore, while the use of stories and narratives in general can facilitate educational outcomes – such as the development of literacy skills (Heath 2004), empathy (Jarvis 2012), memory (Marsh *et al.* 2006), and information sharing (Boyd 2009) – the narrative transportation that occurs through storytelling also reveals some of the unanticipated challenges of internet-enabled teaching and learning throughout the twenty-first century. Narrative transportation suggests that the engaging, immersive experience of a story, which can facilitate strong affective responses and low levels of critical thinking, at the same time can have unintended negative consequences around identity formation, especially across online spaces.[1] Such a notion permeates much of the audio-visual media addressed in this volume.

More positively, however, Kearney (2011) showed that learner-generated digital storytelling often leads to the development of critical thinking, since it allows learners to express personal emotions through autobiographical explorations of a variety of topics and modes of communication (Gretter *et al.* 2017: 12). At least digital storytelling can be part of an active social practice that challenges, contributes and critiques the inequities and ideological discourses embedded within the 'real world'. With the ever-present dangers of greenwashing, however, and taking into account a strategy of simply avoiding the major challenges of climate change, global citizens more than ever need robust and critical modes of media engagement to help face up to these growing concerns. This study fully endorses Gretter *et al.*'s conclusion which cogently affirms that in a time of fake news, with the growth of conspiracy theories and edutainment, the necessity for robust forms of interactive and critical media literacy is more pressing than ever.

Network theory and YouTube

YouTube can serve as a forum for propagating inductive reasoning and new forms of eLearning to the outermost limit across online society, all the while seeking ways to connect marginal, isolated, excluded or just some shy lurking media consumers, so that they may thrive while easily accessing so much material at a click of a mouse. Such a utopian and positive interpretation of its technological affordances portrays humans as 'inductively rational pattern recognisers, who are able to make decisions in ambiguous and fast-changing environments and to learn over time' (Bueinhocker 2006: 138, cited in Hartley 147). This is what YouTube and other new media technology constantly also suggest.

Beinhocker (2006: 141, cited in Hartley 2017) goes on to suggest and even promote that 'networks are an essential ingredient in any complex adaptive system. Without interactions between agents, there can be no complexity'. Finding ways to 'access, understand and create communications in a variety of contexts' remains the everyday work of much new media, which in turn corresponds with how some media regulator defines 'media literacy'. Consequently, as subjects of the new media age, we learn to navigate the 'hierarchies of networks within networks' that characterise both markets in the global economy and meanings in the global sense-making system, including language, the internet and YouTube (Hartley in Burgess and Green 2009: 143). But what literacy skills do people need (to have, to know, to do) in order to participate and learn from YouTube, beyond random copying or contagion, is constantly contrasted with more 'disciplined' and structured teaching.[2] The form and level of literacy needed depends, of course, on the textual affordances of media, as well as the ever-changing user's technical and media capabilities and needs. This chapter seeks to examine how such unique levels of technological adaptability and literacy can be re-purposed to speak to new forms of environmental literacy, as embedded within the YouTube platform, for example.

Do new media formats of communication like YouTube and other internet affordances (open source programming, wikis, blogging, social networks, social bookmark folksonomies and the rest) require investment (public or private) in teaching whole populations how to use them? Or are they better served by embracing ad hoc and blind experimentation and adaptation? Since the arrival of the Graphic User Interface (GUI), for instance, helping to revolutionise computing and 'intuitively' provide effective modes of engagement with computer interfaces using click and past modalities, it has been suggested that new media, and technical computer literacy especially, has become simply intuitive. Consequently, there is no apparent need for formal training, as users simply learn to use such new media protocols through trial and error. Furthermore, it is asserted, or at least assumed, that the same strategy and mode of learning might be applied to all forms of media literacy, including environmental online learning.[3]

Reception of scientific/environmental messages: horizons of expectations

This chapter will examine a selection of popular YouTube output that speaks to a broad range of environmental issues and by extension address literacy agendas, while feeding off potential audience and reception engagement and learning. This connection is set out most directly by the very useful blog by Brigitte Nerlich (available online at: http://blogs.nottingham.ac.uk) 'Science Climate Communication: A View from Reception Theory', which can easily be re-applied to the YouTube phenomenon. She writes: 'Interestingly, as far as I can make out, nobody seems to have written about science communication from the perspective of "reception theory" (reading, interpreting, understanding) aspects of this process.' Reception theory basically deals with 'how people read (or talk, or use images); how they create meaning. It has some roots in good old-fashioned hermeneutics.' The theory also deals with how authors create a space for such interpretation to happen and how they inscribe their imagined readers into the texts they create.

Meanwhile, Hans Robert Jauss invented the term 'horizons of expectation' which has everything to do with a reader's position and nothing to do with projected sunset and sunrise times constituting a landscape, for instance. Essentially, interpretation always interacts with expectations, thereby further extending the 'uses and gratification' model explored in the opening chapters. A pertinent question posed would be how have our 'horizons of expectations' regarding environmental issues in particular, been transformed and extended by the growth of online media and most especially by YouTube. These can incorporate the authors' expectations of how their (intended) reader might understand what they wrote, or the readers' expectations about how to interpret what they are reading or viewing, all the while teasing out their understanding of the author, other texts or images, the genre or style the text is written in, the time they live in, their cultural and political values and identities, and so on. To put it another way, 'horizons of expectations' simply refers to what the reader 'expects' of the text and this will change according to the time and place of the reader. These reception parameters and expectations can be applied directly to a broad range of readings of YouTube videos, which constantly veer between the exigencies of text, medium and users' modes and predispositions.

Like so much analysis of reception theory explored in earlier chapters, uses and gratification theory in particular perceives viewers as active participants in the creation of meaning, rather than as passive receptors. While this approach is probably more applicable to the interactive video games examined in a previous chapter, nonetheless it equally applies to the dialogical nature of the YouTube platform. Depending on social and psychological circumstances, new generational audiences have certain needs and expectations which propel them to seek out, read and engage with certain media and content at any given time. This, in turn, leads to differential patterns of media exposure, which result in the gratification of initial needs, but also has other often unintended consequences. Furthermore, increasing layers of complexity are added in the process, including an element of celebrity-address and particular intertextual knowledge around who is presenting the stories.

Celebrities and YouTube videos

Many young people (from carrying out ad hoc discussions with my students) mention the Hollywood superstar Leonardo DiCaprio as being the most connected with the environmental cause. The actor consistently reiterates how 'climate change is severely impacting the health of our planet and all of its inhabitants', and

> we must transition to a clean energy economy that does not rely on fossil fuels, being the main driver of the global problem. Now is the time to divest and invest to let our world leaders know that we, as individuals and institutions, are taking action to address climate change.[4]

Some hard-core environmentalists take issue with DiCaprio's credentials as an environmental activist, criticising his use of a private airplane and also for engaging in other forms of conspicuous consumption (see Miller 2017). While recognising such legitimate concerns, if such public figures speak passionately and directly to their large audiences and can effectively address the global issues of climate change – as he did in his acceptance speech for Best Actor in the Oscars for *The Revenant* (2015) – then this heavily watched performance remains useful in laying the foundations for more widespread civic engagement with such important concerns. Environmental communication scholars, as suggested throughout this volume, should therefore at least acknowledge the benefits of such interventions and pronouncements. This is a more fruitful strategy than instinctively dismissing them as faulty, by uncovering some hints of an otherwise hidden greenwashed agenda being concocted, or constantly trying to dismiss the celebrity for some otherwise minor slippage of reasoning.

Meanwhile, for an older generation, Arnold Schwarzenegger remains a clear leader and environmental model, having launched the non-profit environmental organisation R20 whose mission is to help sub-national governments around the world develop low-carbon and climate-resilient economic development projects. (See for example environmental readings of his performance in *The Terminator* franchise explored in Brereton 2005.)

> I believe that if we want to fight climate change; if we want a green energy future; if we want real action that matches our vision, we must do a better job of communicating. We have to talk about things that matter to people. We should look at this like a four-legged stool; right now, we're using a one-legged stool', which is wobbly and unbalanced. 'The first leg is jobs; second leg is national security; the third leg is health. Pollution kills. And the fourth leg of course is Climate Change.[5]
>
> (Lanza and Litousky 2012)

Now, in the age of social media, a great opportunity is afforded to the overall communications process, since such simple but effective environmental communication messages can be spread throughout the world. Videos can go viral, and can reach people in all four corners of the Earth. Certainly, such celebrities provide a well-publicised call-to-arms, recognising the pervasive power of media to speak to complex environmental problems which often divide audiences.

Critics of celebrity environmental activism from Schwarzenegger to DiCaprio and more recently Ian Somerhalder, often assume that fans blindly follow celebrities who invite them to support their brand of celebrity activism and charity. Somerhalder has become a celebrity through the cult teen franchise *The Vampire Diaries* (2009–2017) and has used his status also to set up an environmental organisation. Research most recently by Hunting and Hinck examines how such a celebrity can invite fans to use feelings of intimacy to engage in civic action, including environmental activity. They draw on scholars like 'Bennett (2014) and Chouliaraki (2012) to argue that celebrity activism uses feelings for celebrities to mobilize fans in powerful ways.' They further suggest, there are three elements used in the process of engaging with 'on-screen characters' personas and narratives, which in turn 'can be knitted together in ways that not only make a particular celebrity culturally meaningful, but in ways that can drive civic action through celebrity activism' (Hunting and Hinck 2017: 433).

Meanwhile, work by Brockington and Henson (2014) asserts that celebrity 'is now part of the way that most major charities and particularly environmental development charities, go about raising funds, raising awareness and lobbying for their causes' (in Hunting and Hinck 2017: 432). Rojek (2001) asserts that 'one particular tension in celebrity is that the arousal of strong emotion is attained despite the absence of direct, personal reciprocity' (12). Fan studies scholars for instance calls these extensive feelings of intimacy, emotional investment, and commitment 'affective identification' (Hills 2002; Van Zoonen 2005).

Chouliaraki (2012) explains that this kind of celebrity activism mixed with humanitarianism, 'prioritizes the "authentic" emotions of the celebrity and our own connectivity towards her' (17). Through celebrity endorsement across so many YouTube videos, together with more directly addressed climate-focused music videos, we can feel outrage at extreme suffering on the other side of the world. Critics of celebrity activism, however, assert that such forms of celebrity activism lack substance and alternatively do a better job of strengthening fans' commitment to celebrities than their on-the-ground commitments to complex public issues. Internet studies and celebrity scholars have demonstrated that social media has intensified celebrities' performances and fans' expectations for intimacy with their online heroes.[6]

Leonardo DiCaprio: *Before the Flood*

*D*irected by Fisher Stevens and starring Leonardo DiCaprio, it was described by *National Geographic* as invoking the power of 'fine art' to influence an analysis of climate change. The documentary, which is constantly mentioned by students, follows the celebrity actor as UN climate ambassador on his journey for two years, leading up to speaking at a New York Climate Meeting to ratify the Paris agreement. The apparent spark or creative imaginary and deep memory for

Leonardo's ongoing fixation with environmental issues, emanates from being exposed to a historical painting 'The Garden of Earthly Delights' (1500) by Hieronymus Bosch, which somewhat strangely was hanging on his crib as a child. The painting constituting an infamous triptych, with panels showing an idealised Garden of Eden and representations of Adam and Eve on the left, alongside with what an apocalypse might look like in the centre, cataloguing millennial fears of an unstable age of damnation for all types of sins. Finally, on the right is a powerful and dramatic image of hell. Especially troubling and worrying, this historical painting displays a dystopian fantasy of what could happen to us (and our planet) through such an image of hell. Representing a dystopic and cautionary horror tale of its age, this provocative two-dimensional artwork with three contrasting visions had a profound effect on the child-star and has stayed with him all his life.

One hopes some new media audio-visual artefacts could become as powerful and provocative for new generational secular audiences in the future. DiCaprio's memories, and others' audio-visual environmental work, while noting that contemporary viewers, do not appear to subscribe to the power and influence of such Old-Testament meta-narrative fables.[7] At one stage in discussion, the celebrity star announces, 'if this was a movie, we would solve the problem'. One might again wish such a dynamic relationship between information and achieving a subsequent solution were possible. The charismatic actor continues: 'it's up to us – we will be judged by what we leave behind'. Like all good populist documentaries with a clear educational trajectory, the story-line provides some simple yet powerful take-away environmental messages to entice the viewer, including:

- Consume less meat (and in turn watch *our* excessive consumption patterns).
- The documentary particularly highlights a major environmental issue concerning Palm Oil – causing so many forests in Sumatra and Indonesia being felled – which in turn results in excessive amounts of CO_2 escaping into the atmosphere. Unfortunately, this relatively cheap produce is used as an additive in everything from Burger King products to crisps. Consequently, the public needs to be made aware of such unfortunate imbalances in the global material distribution system and documentaries like this one call attention to this educative process.
- We are addicted to cheap energy, which remains the major dilemma of our age.

Environmentalists have been struggling with ideas around how to reverse this tendency and to help strive towards promoting a low carbon energy future. For instance, the documentary calls attention to the innovative technology company Tesla, which promotes a technological solution, of course. This call on big capitalism for environmental solutions remains an obvious, but often precarious option, as discussed across several chapters in this volume. Apparently, the documentary suggests 'we' can build 1,000 of these 'super' battery-powered plants that will solve the energy needs of the whole planet. Primarily, the documentary posits how society should keep carbon in the ground, which is necessary for the future sustainability of the planet.

Addressing key concerns regarding climate justice and the Third World. DiCaprio talked with the Roman Catholic Pope Francis regarding justice and climate change and symbolically exchanged presents of books, including a reproduction of the celebrated painting discussed above, alongside speaking directly with other celebrity politicians including President Obama. Hence, the actor's celebrity status as 'global mover and shaker' becomes further solidified and re-affirmed.

Throughout the documentary there are some insightful images displayed of scientists helping audiences to see the planet's thin membrane through the upper atmosphere and calling attention to massive (beautiful and artistic) simulations of how the planet will get dryer in the middle, while affirming there will be more rain in the north, coupled with a much colder Europe. All of these dramatic climate change predictions will, of course, cause major problems for the planet. More dramatically, the planetary destruction will begin with small low-lying islands, where all forms of life will be totally lost, as they become submerged in the sea. In particular, natural eco-systems like the Great Barrier Reef will be decimated by up to 50% in less than 30 years. The documentary follows a conventionally alarmist path. Such fertile eco-systems have been essential towards feeding over one billion people, leaving, of course, huge concerns for food security into the future. As is constantly pronounced by environmentalists and other political thinkers, 'we just don't think long-term'.

While recognising the danger of preaching to his audience, because DiCaprio is a bona fides star actor, he can get away with various modes of critical address, which sometimes involves pontificating. It is what actors do professionally and consequently he can be considered above the cut and thrust of (grubby) politics. Therefore, when such artists and creative communicators stay inside their protective mask, while producing powerful media performances that directly speak to environmental issues, their resultant environmental messages become even more provocative and engaging. This is evident in the following section, which drills down into the creative work of a number of music bands like Radiohead and most especially recalling pop singers including Michael Jackson, Julien Lennon, Bjork and, particularly for the purposes of this chapter, Prince Ea, all of whom at times devote their creative skills towards calling attention to and promoting a broad range of environmental concerns.

Environmentally themed music video on YouTube

Drawing on the political economy of the music industry, technological change and not-unrelated economic developments challenge common sense assumptions about media forms and practices. Historian William Uricchio has suggested that the processes of digitisation and convergence are redefining 'our present as a moment of media in transition' (cited in Wurtzler 2007: 279).

According to Steve Wurtzler, virtually any new form of mass media has been hailed as the instrumental fulfilment of a perpetually deferred participatory public sphere. First, of course, we had television, then cable and more recently the internet, which has been broadly made available in our homes, across culture, education, news etc., while facilitating individuals' inclusion in a broader social fabric and supporting informed participation in public affairs.

All the while, the contemporary cultural moment takes for granted the notion of 'spin', be it music or video or both, as we assume a priori that all messages reaching us through their various mediations are shaped and massaged by the sender and most specifically by the industrial apparatus. Yet at the same time, many communication scholars assert that audiences culturally long to experience the unmediated, the direct and the transparent. Wurtzler, for instance, believes that blogging relies on a technology of mass mediation to make the process direct and uncomplicated. Can the same be asserted for other new media platforms and formats, including YouTube? There remains the recurring assumption that technological change offers us access to mainstream grand utopian promises.

Despite being the most widespread form, there is a paucity of scholarship on climate change vis-à-vis YouTube, much less highlighting the culture of popular music (see Ingram 2010). For Andrew Revkin, for instance, narrativity in music can be considered a mode of 'conveying stories about the environment' (Wodak 2017: 1). Meanwhile Georgina Endfield and Carol Morris suggest that, by the beginning of the twenty-first century, climate change has become the predominant environmental narrative, recalling the importance and power of everyday popular music to speak to such concerns. The phrase 'music is the soundtrack of our lives' was coined by Dick Clark (1994) when he declared that 'popular music is the soundtrack of our individual lives. Anything that ever happened to you, good or bad, was scored with the music you listened to' (Endfield and Morris, and Clark, cited in Wodak 2017: 2).[8]

In uncovering historical examples of direct environmental engagement, one immediately comes upon two songs by two highly commercially and critically influential Generation X pop musicians: *Saltwater* by a 28-year-old Julian Lennon from 1991, and *Earth Song* by a 31-year-old Michael Jackson recorded back in 1989. Both songs can be seen to illustrate topical environmental themes of the period, including ozone depletion, deforestation and species extinction. As with Bob Dylan's earlier generational *The Times They are A-Changin'*[9] or Joni Mitchell's *Big Yellow Taxi*, such songs situate the shifting baselines of contemporary timelines, relative to a prior biophysical baseline of a pristine 'nature' that has not been anthropogenically perturbed. The song and subsequent video for *Saltwater*, for instance, suggestively conveys a mournful response to the relatively degraded bio-physical baseline occasioned by industrial civilisation, wherein Lennon intones that 'Time is not a friend/As friends we're out of time', to highlight and redress the rapidity of environmental destruction (Wodak 2017: 4).

Saltwater and *Earth Song* have received over one million and 147 million views respectively on YouTube (as of 1 September 2017) in addition to considerably more original digital sales/streams and hard copy sales. There are over 15.5

million copies sold for *Earth Song* – the third most of any singles by a male artist. Yet despite such outreach and influential music and videos being produced, sustained cultural and media studies on their particular approach to narratology have not been considered, especially with regards to communicating climate change (Machin 2010).

Michael Jackson's *Earth Song* is of course a call to save the planet from the destructive impact that has been wrought upon the earth by humanity and technology. The song was recorded in 1995, but never released as a single in the US, due to events related to allegations around Michael Jackson's private life. Nevertheless, it won a Grammy nomination in 1997 and earlier recognition in the form of a Genesis Award in 1996. *Earth Song* has a specific synchronisation of semiotic modes, orchestrated along four narrative strands and filmed in four different geographic locations across the globe, presumably occurring at the same time. Each of these strands presents images of deforestation, animal cruelty, pollution and war, with their disastrous consequences for humanity and Earth.

Popular music most effectively operates at the interface between the tensions embedded within climate-as-science, alongside being constantly rooted in its universal cultural significance. Writing from the perspective of historical geography, Richard Hamblyn suggests that: 'all narratives of climate change deal inescapably with uncertainty, whether they are supportive of the consensus scientific view or not, while detailed scientific claims and counter-claims only add to the sense of confusion apparently felt by lay audiences' (2009: 224, cited in Wodak 2017: 5). In varying ways, environmentally themed music and music videos can certainly help dispel such confusion and actively promote fruitful engagement across a range of cogent environmental debates.[10]

For instance, it is argued that narrativity in music can be considered a mode of overarching narratives 'conveying stories' rather than 'communicating' them. This pervasive strategy calls for the adoption of the 'creative imaginary' (Brereton 2016) of sound and music to synthesise and aid in this overall communication process, suggesting in turn a major paradigm shift is required in the communication of climate change issues across mass media communication. The former approach denotes an idea, impression, or feeling as distinct from the latter that denotes information or news. Music when it most explicitly engages with climate change and the tensions inherent in such discourses, sits at the communication rather than the conveyance of information end of the spectrum.[11]

The power of direct address: Dear future Generation – Sorry: *Dear Future Generation* by Rapper Prince Ea (April 2015)

An analysis of Prince Ea's video which has reached over 9.6m views, helps to frame the importance of celebrity in getting across such environmental messages. The rap music video and elegiac, emotional direct-address at the audience, begins with the image of a desert, as the celebrity poet/rapper, dressed in a conservative suit, walks into frame and recounts his personal oral story, all of which is unadorned and spoken direct to camera. Recalling the Bardic function of television, as suggested by John Hartley and using the hypnotic power of direct address, YouTube most certainly has co-opted this otherwise unsophisticated form and style to great effect. Visual highlights of the video, include the use of words on the screen to reinforce what is being said and to dramatise its importance, both as rap poetry and recalling its ideological and environmental meaning. All of which underpins the intergenerational angst bubbling up from not facing up to, or actively dealing with the global issues around climate change. By calling on the emotive power of injustice, a key strategy sets out to emphasise how current generations, who are made aware of the problem appear to do nothing, primarily because it will *only* affect future generations. Such short-termism, unfortunately, will be most keenly experienced by new generational audiences in years to come and hence the celebrity rapper actively appeals to cross-generational concerns for justice and the long-term fall-out, if such global environmental concerns are not addressed and urgently.

Most memorable recorded in the video is the use of white papier-mâché images of trees,[12] which serve as a short-hand signifier and barometer of the precariousness of environmental health. Real organic trees, growing in their natural habitat are needlessly destroyed, inferring all the while that trees remain the fountain of all life on the planet. Later, recalling the similar use of a close-up of the dollar bill that was used in a famous revenge episode from *The Sopranos* explored in Chapter 7, the dollar bill most directly serves as a short-hand signifier for consumer capitalism that is quickly destroying the planet. Prince Ea certainly gets across his questioning of excessive materialism and its effects on the planet in a more powerful and direct manner that effectively speaks to new generational audiences. Within the format of an elemental music video, the combined narrative trajectory of the story helps to symbolise the commodification of everything of real value on the planet. Simply and most pointedly, the celebrity rapper concludes his rap with the entreaty that 'we can't eat money and it will not sustain us'.

While critically it would be relatively easy to reinforce a totally fatalistic message of despair to close this emotional plea for transformation, instead the audience is afforded real hope and direction for change. Such a positive communications strategy remains incumbent on artists as well as educators, wishing to constructively question reality, while highlighting the lack of progress in tracking such environmental issues. At the same time this posits alternative solutions and signalls provocative scenarios that might inspire audiences to become active in this ongoing struggle themselves. Enabling audiences to take control, while providing a coherent environmental world-view, remains a constant refrain across such popular videos. What more can a music video do – or any work of art for that matter – but actively engage with mass audience's emotional and empathetic connection with nature?[13]

In another of his popular videos, *Will this be Humanity's fate*, Prince Ea begins by telling a story about picking up his nine-year-old nephew from school. Out of the blue, the boy asks, why is humanity killing the planet? His rapper uncle explains using 'simple graphics', again showing how trees are necessary for life, but illustrating how these life forms are being killed by factories spewing out their toxic waste. 'The earth is a living organism, if it dies, we die' – recalling the ongoing planetary trajectory of over-consumption. Why are *we* so fixated with 'smart machines', becomes the continuing refrain, while 'smart humans seem to be in short supply?' The rapper further repeats very memorable and catchy refrains, including: where can I sign up to be 'primitive' or recalling the well-worn cliché, 'why do we choose greed over need'?

The singer goes on to add colourful and poetic aphorisms such as, 'it's too late to be pessimistic'. All the while calling humans out for a growing need to face up to *our* responsibilities.

Most notably in another video, *Why I'm happy Trump Won*, he insightfully explores how people don't necessarily change when they are comfortable: 'it's time he says to wake up from our deep illusions.' Facing up to (environmental) conflict by abdicating political responsibility to elected leaders is certainly not enough within current turbulent times. This is a response echoed by Al Gore in discussions around his *An Inconvenient Sequel* discussed in Chapter 5. Memorably, in calling attention to the allegorical conceit of the Rubik's Cube, the rapper suggests that *we* solve problems only when all elements of the matrix are in alignment. Such holistic problem solving remains an abiding productive approach of climate change in particular.

Certainly, one could characterise Prince Ea as being somewhat overly idealistic, recalling dreams of a transformational utopia. Those folks just weren't listening.

> All he is asking us to do however is to look up from our screens, just long enough to rethink our choices, just a little bit, and make an effort to question the status quo, as well as just be kind – it isn't that outlandish of a request.
>
> (Cundiff 2013)

All of this subtle engaging rhetorical verbiage and call to individual action remains very far removed from the usual, if dated, criticism of rap music – as being destructive and misogynistic, while celebrating violent criminal lives that both embrace a gun culture, together with a deviant life-style involving drug taking.

Of course, conventional media analysis, recalling the long-established cultivation theory in particular, is often used to highlight the negative effects of crude and often pernicious rap lyrics cited above on susceptible students/listeners attitudes. But such an approach can also be more usefully turned on its head with regard to the active promotion of pro-social messages. Pro-social and environmental sentiments embodied in these YouTube videos can support the propagation of effective environmental messages, or at least sow the seed of alternative and more productive engagement with such important planetary issues.

For instance, Hip-Hop culture and rap music (see, for instance, Raphael Travis Jr) was born through an environment that was at once both oppressive and yet at the same time unyieldingly innovative. Present day rap music shares these characteristics, despite existing within a different landscape. In current times, rap music weaves throughout the fabric of pop-culture and uses new forms of poetry and imagery to speak to issues around injustice and oppression, as much as environments of unbridled wealth and prosperity. Furthermore, at the other end of the spectrum, consumers of contemporary rap music, and even less the growing YouTube audience, do not fit neatly into stereotypical demographic genres either. We find consumers across all ages, alongside racial and ethnic divisions, gender and geographic regions, actively engaging with the pleasures of music and taking into consideration its sometimes opaque and explicit messages. Hip-Hop incidentally enjoys as much praise in certain circles as it is maligned in others (Rose 2008). By any measure the YouTube platform for reconstituting such musical idioms opens its appeal out to a global audience, which is not restricted either spatially or temporarily in any way, to fit into prescribed ethical, socio-economic groupings or other divides.

Somewhat paradoxically, the profit-driven commercialisation of the ever-growing popular musical culture is as much responsible for its proliferation, and maybe even more so according to some commentators, than is the desire for artists to push the boundaries of creativity and self-expression. By any measure, such online technology makes these musical idioms and their enticing and radical messages, especially around environmental issues, even more widely available and accessible. Surely this alone affords hope for the future of environmental media.[14]

Concluding remarks

See, for example, the conceptual model of empowerment, which is necessary to speak to audiences through environmental and other issues and is based on positive youth development, as adapted from Travis (and Leach) (2012). At the top of their pyramid is 'engaged citizenship', with a sense of community underneath, supported by confidence and competence. Under this rubric is a focus on connection, flanked by character and caring attributes (2012: 160). In a world that is becoming smaller through globalisation, and most particularly recognising the common and shared cultural benefits of a universal online media platform like YouTube, it is vitally important to the long-term sustainability of humanity that new generational youth develop into healthier and more significantly engaged contributors to the environmental betterment of all our communities. At the same time we must recognise a steadily aging population across more developed countries, where global power remains most concentrated. One can certainly agree that musical genres like Rap, Pop and Hip Hop together with Rock, Folk and Country displaying their emotive use of pro-social environmental messages can assist in this process. Explicitly conceived environmental Rap videos, like *Sorry* in particular, one might hope are consumed by mainstream new generational audiences for the direct environmental messages they afford.[15]

The enormous growth of YouTube and its constituent elements, especially supporting short promos and other video streaming services like Netflix, needs to be constantly evaluated, especially noting how such platforms proliferate across the online mediaspheres, spaces that new generational audiences occupy. Such new forms of media output ought to become an ongoing focus of environmental scholarship and audience research. I would like to believe that these new and exciting media artefacts speak directly to a musically inspired YouTube audience. Such musical promos and audio-visual texts are by any measure not narrowly addressed to climate aware citizens, but alternatively call on widespread attention across generations to our understanding and communication of the challenges faced by climate change.

Innovative new strategies are always needed to speak to mass audiences, calling upon progressively aesthetic formats and alternate voices like those illustrated in this chapter. We certainly need more and more creative imaginaries to get the environmental message(s) across, beyond the clichés of polar bears on ice-flows and factories emitting toxic fumes. The powerful musical constituency of Rap lyrics, as illustrated by Prince Ea, for instance, alongside the huge appeal of celebrity culture online, can go some way in communicating important messages. Adopting non-preaching modes of storytelling need to be constantly deployed in new and innovative ways, as re-constituted media formats and online platforms are deployed addressing climate change while acknowledging that traditional aesthetic strategies, and even the use of stereotypical images, continue to have a place in communicating environmental messages. Even if we wished, there is no magic bullet formula, much less a clearly defined monolingual way to communicate such a multi-layered confluence of environmental issues. Concurrently, environmental artists and communication scholars constantly need to explore new and innovative ways that connect with the accelerating affordances of these ever-changing media platforms and outlets that can help address the global challenge of our time.

Such debates and strategies of engagement need to speak to more creative imaginaries. Drawing on research that speaks to a range of environmental problems are constantly required, beyond the short-term use of iconic signifiers like Prince Ea's paper trees. As educationalists we need to learn from pedagogical best practices across other forms of educational and (new) mass media literacy protocols, to help achieve a low carbon future. We certainly need more practical and bespoke communication exemplars to be developed to assist in formulating new forms of eco-literacy that connect with a broad range of publics and citizens across the world. Looking at how YouTube short films and music videos impact on young people's pleasures and storytelling protocols remains a useful example. Relating to the particular affordances of such new media formats remains an important first step in understanding and creating active prosumers and even support future critical environmental media practice.

Coda: facing up to the emergency

Despite the scientific community's agreement on the anthropogenic roots of the current environmental state, it would appear that explicit manifestations of climate denial have unfortunately grown stronger. In this context, Suzanna Priest is correct in stating, 'climate change is something of a communication emergency' (2016: 9). She observes that the field is at a crossroads, characterised by a move from an information deficit model, which involves a top down transmitting of information, which was thought to persuade individuals to change their behaviour, to a dialogical model of communication that emphasises effective and more long-term public engagement between scientists and non-scientists (the public). This paradigm shift from simple information provision and persuasion, to more active engagement, is however far from insinuating a separation between individual and collective approaches (Castro-Sotomayor 2017: 1).

One could hopefully conclude that the use of persuasive rap-poetry like *Sorry* on YouTube straddles both trajectories and is certainly not afraid to provide information in a persuasive and convincing manner. While putting most of its effort into attempting to convince its younger audiences of the need to face up to climate change, such an evocative video uses a fresh range of cognitive, emotional and aesthetic strategies around the need to actively engage with environmental problems. We need to support more effective environmental media artefacts that speak directly to audiences who are not necessarily predisposed to an environmental message.

Notes

1 Fanfiction, online advertising and radicalisation provide examples of how young individuals face challenges in determining the blurred line between facts and fiction. While these challenges are significant, educators should not shy away from using the affordances of digital storytelling for teaching (Gretter *et al.* 2017: 10).
2 And what might be expected if 'we' – the users – decided to make a platform like YouTube useful not just for self-expression and communication but for description and argumentation too – for 'objective' as well as 'subjective' knowledge, in Karl Popper's terms (see Kewth 2005).
3 Can we imagine a hybrid formal/informal (expert/amateur; public/private) mode of propagation for learning 'digital [environmental?] literacy', and if so, how might YouTube in particular play a part? (Hartley in Burgess and Green 2009: 128). More interestingly from an educational and literacy perspective, John Hartley wonders how might YouTube be exploited for scientific, journalistic and imaginative purposes, as well as for self-expression, communication, and file-sharing (ibid.: 128–129).
4 Available online at: http://fortune.com/2015/09/23/leonardo-dicaprio-fossil-fuels/.
5 See 2012 Alliance interview with the star-environmentalist. Available online at: www.alliancemagazine.org/feature/interview-arnold-schwarzenegger/.
6 'Somerhalder enacts in his social media posts with fans and the same values he enacts in his performance of his on-screen character, which in turn can lead to and most certainly invite slippage' (Hunting and Hinck 2017: 442).
7 The paintings recall an Adamite heretical sect from the period, who believed the world was coming to an end. Note, before the global disaster of the flood, people who were primarily vegetarian suggesting a 'healthy organic state of nature', had to change

because of a lack of vegetation to become meat eaters, together with other more deviant behaviour patterns.

8 In a subsequent article written with Carol Morris, Endfield argues that this data-centric climate change narrative has given rise to the global 'meta-narrative' being 'the predominantly scientific discourse in which this is couched, and the increasingly global-scale of climate thinking' (Wodak 2017: 4).

9 See also the photographic project using Bob Dylan's anthem *Hard Rain* (2006) which was re-launched through the *Whole Earth* (based on the premise that the future belongs to today's young people) exhibition seen in numerous universities around the world since 2015. The Hard Rain Project. Available online at: www.hardrainproject.com

10 The planetary scale profundity of the existential dilemmas that are encapsulated by the Anthropocene mean that popular cultural engagement is even more pertinent than that with climate change. However, the proposal to rename the current geological age is relatively recent and is only just starting to permeate wider society (Wodak 2017: 5).

11 For instance, the ubiquitous representation of climate change via Cartesian graphs of J-curves has a low signal:noise ratio. Yet this communicative approach is antithetical to wider societal engagement, as it privileges intellectual representation over emotive engagement with those 'subjects, situations and feelings' that 'cry out' in response to climate change.

12 Recalling the iconic imagery in the original *Blade Runner* and the paper images which displayed the replicants power of 'creative imaginary' (see environmental reading by Brereton 2005).

13 This especially applies to the complex dilemmas around climate change, all the while speaking directly to both the emotional and cognitive intelligence of their audiences.

14 Many web and satellite-based mechanisms with MP3 enabled devices essentially make music 'free' and portable including YouTube, Spotify, iPhones, iPads and many other file sharing applications.

15 I have, for example, shown this climate change promo *Sorry* (https://youtu.be/eRLJscAlk1M) to over 200 first year students as part of a series of provocative online sources on our Communications/media courses and they have responded very well to such stimuli.

References

Bennett, L. 'Tracing Textual Poachers: Reflections on the Development of Fan Studies and Digital Fandom' *The Journal of Fandom Studies*, 2(1): 5–20. 2014.

Bott, C. *The Internet as a Terrorist Tool for Recruitment & Radicalization of Youth*. DHS report. 2009. Available online at: www.homelandsecurity.org/docs/reports/Internet_Radicalization.pdf.

Boyd, B. *On the Origin of Stories: Evolution, Cognition, and Fiction*. Cambridge, MA: Harvard University Press. 2009.

Brereton, P. *Hollywood Utopia: Ecology in Contemporary American Cinema*. Bristol UK: Intellect. 2005

Burgess, J. and Green, J. *YouTube: Online Video and Participatory Culture*. Cambridge UK: Polity Press. 2009.

Castro-Sotomayor, J. '"Review" of *Communicating Climate Change: The Way Forward* by Susanna Priest' *Language and Ecology*. 2017.

Chouliaraki, L. *The Ironic Spectator: Solidarity in the Age of Post-Humanitarianism*. Cambridge: Polity Press. 2012.

Costanza, R., d'Arge, R., de Groot, R., Farber, S., Grasso, M., Hannon, B., Limburg, K., Naeem, S., O'Neil, R., Paruelo, R. G., Sutton, P. and van den Beldt, M. 'The Value of the World's Ecosystem Services and Natural Capital' *Nature*, 387: 253–260. 1997.

Cundiff, G. 'The Influence of Rap and Hip-Hop Music: An Analysis of audience Perceptions of Misogynistic Lyrics' *Elon Journal of Undergraduate Research in Communications*, 4(1): 1–41. 2013.

Downs, A. 'Up and Down with Ecology – the "Issue-attention Cycle"' *Public Interest*, 28: 38–50. 1972.

Fletcher, R. 'Nature is a Nice Place to Save but I Wouldn't Want to Live There: Environmental Education and the Ecotourist Gaze' *Environmental Education Research*, 21(3): 338–350. 2015.

Green, J. and Burgess, J. *YouTube: Digital Media and Society Series.* Cambridge UK: Polity Press. 2009.

Green, M., Chatham, C. and Sestir, M. 'Emotion and Transportation into Fact and Fiction' *Scientific Study of Literature*, 2(10): 37–59. 2012.

Gretter, S., Yadau, A. and Gleason, B. 'Walking the Line between Reality and Fiction Online Spaces: Understanding the Effects of Narrative Transformation' *Journal of Literacy Education*, 9(1): 1–21. 2017.

Hartley, J. *The Uses of Digital Literacy*. London: Routledge. 2017.

Heath, S. B. 'What no Bedtime Story Means' *Literacy: Major Themes in Education* 1: 168. 2004.

Hills, M. *Fan Cultures*. London: Routledge. 2002.

Hunting, K. and Hinck, A. 'I'll See You in Mystic Falls: Intimacy, Feelings, and Public Issues in Ian Somerhalder's Celebrity Activism' *Critical Studies in Media Communication*, 34(5): 432–448. 2017.

Ingram, D. *The Juke Box in the Garden: Ecocriticism and American Popular Music*. Amsterdam: Rodopi. 2010.

Jarvis, C. 'Fiction, Empathy and Lifelong Learning' *International Journal of Lifelong Education*, 31(6): 743–758. 2012.

Jauss, H. R. *Toward An Aesthetic of Reception*. Minneapolis: University of Minnesota Press. 1982.

Kearney, M. 'A Learning Design for Student-generated Digital Storytelling' *Learning Media and Technology*. 2011.

Kewth, H. *The Philosophy of Karl Popper*. Cambridge: Cambridge University Press. 2005.

Krajewski, J. M. T., Schumacher, A. and Dalrymple, K. 'Just Turn Off the Faucet: A Content Analysis of PSAs about the Global Water Crisis on YouTube' *Environmental Communication*. 2017.

Kyong, C. and Hui, W. *Updating to Remain the Same: Habitual New Media*. Cambridge MA: MIT Press. 2016.

Lanza, D. and Litousky, A. 'Interview – Arnold Schwarzenegger' *Alliance* magazine. 2012. Available online at: www.alliancemagazine.org/feature/interview-arnold-schwarzenegger/.

Lundby, K. (ed.) *Digital Storytelling, Mediatized Stories: Self-Representations in New Media*. London: Peter Lang. 2008.

Machin, D. *Analysing Popular Music: Image, Sound, Text*. London: Sage. 2010.

Mailer, A., Maier, Carmen D. and Cross, J. L. 'Multimodal Analysis of the Environment Beat in a Music Video'. In Djonov, E. and Zhao S. (eds) *Critical Multimodal Studies of Popular Discourse*, pp. 109–124. London: Routledge. 2013.

Marsh, E. J., Meade, M. L. and Roediger III, H. L. 'Learning Facts from Fiction'. *Journal of Memory and Language*, 49(4): 519–536. 2006.

Millard, E. 'Writing of Heroes and Villains: Fusing Children's Knowledge about Popular Fantasy Texts with School-based Literacy Requirements'. In J. Evans (ed.) *Literacy Moves On*. Portsmouth, NH: Heinemann. 2005.

Miller, T. *Greenwashing Culture.* London: Routledge. 2017.

Murphy, P. *The Media Commons: Globalisation and Environmental Discourses.* Champaign IL: University of Illinois Press. 2017.

Priest, S. *Communicating Climate Change: The Path Forward.* Basingstoke: Palgrave. 2016.

Reid, M. 'Writing Film: Making Inferences when Viewing and Reading' *Reading*, 37(3): 111–115. 2003.

Rideout, V. 'The Commonsense Census: Media Use by Tweens and Teens: Common Sense Media Research'. 2015. Available online at: www.commonsensemedia.org/research/the-common-sense-census-media-use-by-tweens-and-teens.

Rose, T. *The Hip Hop Wars: What We Talk About When We Talk About Hip Hop – And Why It Matters*. New York: Basic Books. 2008.

Shapiro, Matthew and Park, Han Woo. 'Climate Change and YouTube: Deliberation Potential in Post-video Discussions' *Environmental Communication.* 2011.

Taylor, Pamela. 'Press Pause: Critically Contextualising Music Video in Visual Culture and Art Education' *Studies in Art Education*, 48(3): 230–246. 2007.

Travis, R. 'Rap Music and the Empowerment of Today's Youth: Evidence in Everyday Music Listening, Music Therapy and Commercial Rap Music' *Child and Adolescent Social Work Journal*, 30(2): 139–167. 2012.

Van Zoonen, L. *Entertaining the Citizen: When Politics and Popular Culture Converge.* Lanham MA: Rowman and Littlefield. 2005.

Wodak, Josh. 'Shifting Boundaries: Conveying Climate Change in Popular Music' *Environmental Communication*, 12(91): 58–70. 2017.

Wurtzler, Steve. *Electric Sounds: Technological Change and the Rise of Corporate Media.* New York: Columbia University Press. 2007.

Zelin, Aaron Y. 'Picture or it Didn't Happen: A Snapshot of the Islamic State's Official Media Output' *Perspectives on Terrorism* 9(4): 85–97. 2015. Available online at: www.terrorismanalysts.com/pt/index.php/pot/article/.

11 Conclusion

Constructing an environmental literacy consensus through new media

Overview: narration and new generic modes as a focus for environmentalism

Alex Law argues that

> [y]oung people in particular are thought to be susceptible to self-conforming networks. They engage far less with traditional news consumption and are easily bored by political reporting. Social media reinforces a culture of distraction among young people instantly feeding on depthless fragments of information.
>
> (2017: 6)

While many scholars appear to agree with Law's assertions, nonetheless new generations could also be exposed to a wider range of viewpoints, which this volume seeks to illustrate. Evolving modes of new digital narrative and generic formats, coupled with an understanding of audiences needs and trigger points, highlight some key aspects of this study's investigation into environmental literacy. Focusing on a textual reading of creative imaginaries that speak to the complexity of environmental issues and debates as encapsulated by the spectre of climate change, has been the focus of this broad survey of conventional and new media formats and platforms. While each chapter sets up its own terms of reference with some tentative conclusions, this final chapter will tease out a few more salient points for further discussion and provide some take-away observations.

For instance, empirical research from the 'Stanford History Education Group' recently showed that out of 7,804 student responses, more than 80% of middle school students believed that web ads were real news stories and more than 80% of American high school students had a hard time distinguishing between real and fake photos. The large study concluded that, 'overall, young people's ability to reason about the information on the Internet can be summed up in one word: bleak' (2016: 4, cited in Gretter *et al.* 2017: 2). While probably over-stretching the registering of well-honed moral panics around the divisive power of new media, one wonders if the reality is any better within Higher Education across the Western world – or across the general populace for that matter – with regards to appreciating the overall complexity of climate change, most importantly in developing and promoting robust forms of environmental educational together with effective media coverage and engagement.

Certainly, recent contentions about so-called fake news and misinformation online, according to Gretter *et al.* (2017) has shed light on the critical need for innovative and new forms of media and environmental literacy, which is further evident within climate change discourses. Using contemporary examples from news stories, fanfiction, advertising and radicalisation, Gretter outlines the unique features, affordances and real-life implications, for such media engagement, across a range of interdependencies facilitated by a broad range of digital stories.

Gretter *et al.* convincingly argue that an essential step in teaching students to critically assess online content is to first understand how the human mind distinguishes between reality and fiction, when being transported through online storytelling spaces. Most specifically for this study, among other questions, we have asked: how do digital stories impact youth's sense of reality? Hopefully, this overview survey advances a step in this direction by interrogating online (fictional) stories through the lens of narrative transportation theory. As outlined in earlier chapters, we have tried to uncover: (1) what makes stories a powerful tool to convey ideas; (2) how digital storytelling differs from traditional storytelling; and (3) what educators can do to help students critically assess stories in online spaces (Gretter *et al.* 2017: 2). This is a process which directly dovetails with this book's examination of various forms of new media narratives, exploring their potentially beneficial import and unique affordances towards promoting various aspects of effective environmental learning and engagement.

Storytelling in the twenty-first century has evolved as people find new ways to record, share and consume stories – including games, personal experiences, or news among others (Lundby 2008). Today, modern technology has merged the visual, auditory and the textual, while giving everyone the potential to become their own storyteller and make use of their ever expanding digital environment for personal purposes (Greenfield 2015).[1] Other scholars remind us that psychological immersion into someone else's story – also called 'narrative transportation' – entails a development of emotionality and developing an attentional focus, which becomes a very powerful tool in effectively communicating with audiences. Several persuasive research studies have suggested that when readers' pre-reading emotional states match the emotional tone of a narrative, transportation into that particular narrative is greatly increased (see Gretter *et al.* 2017: 3). Consequently, it makes sense that the mediated message has to both cognitively and emotionally connect with its designated audience to become fully effective.

Through stories we share fragments of life and communicate with others for multiple purposes – such as imparting knowledge, leaving a personal legacy, or simply using them for entertainment (Zipes 2013). In fact, as human beings, we naturally think in narrative ways, which sometimes makes it difficult to realise how central stories are to our human experience (Turner 1996). Transportation into a narrative world is essentially an experience that involves affective responses to powerful imagery (Green and Brock 2000: 703).[2] However, in the current mixed media landscape, scholars can further question if narrative transportation functions differently, or not as the case may be, in an age where stories are widely shared and sometimes repurposed online. Gibson (1979) explains ways that human beings make use of the affordances that a specific environment offers them, while Jenkins (2010) called such a phenomenon 'transmedia storytelling', involving popular stories generating a variety of cross-media adaptations (Gretter *et al.* 2017: 6).

By all accounts, the ubiquitous power of the internet can sometimes blur the boundaries between reality and fiction, rendering the distinction between factual and fictional information more difficult (Fuchs 2007) – a phenomenon, which is illustrated across several chapters in this study. Meanwhile, one wonders what media formats and genres can be most effectively deployed to promote constructive forms of environmental literacy, while at the same time critically learning the lessons of the inherent dangers of sensationalist hype and (greenwashed) PR. This tension is illustrated by a critical reading of *James May's Bright Ideas* in Chapter 5, alongside a case study of an Irish farming advertising campaign 'Origin Green' explored in Chapter 3.

As also evidenced in this study, social networks and archive digital fora coupled with ever new forms of digital software are used to disseminate innovative forms of videos, computer games, music clips, memes or photos (Bott 2009). This can often rely more on visuals, rather than text-based narratives to help relay emotive media messages.[3] Furthermore, while the use of stories and narratives in general can facilitate educational outcomes – such as the development of literacy skills (Heath 2004), empathy (Jarvis 2012), memory (Marsh *et al.* 2006), and information sharing (Boyd 2009), alongside environmental literacy, the narrative transportation that occurs through storytelling also reveals some of the unanticipated challenges of internet-enabled teaching and learning throughout the twenty-first century. Yet by all accounts there certainly is no magic bullet formula for effective pedagogical practice.

Narrative transportation suggests that the engaging, immersive experience of a story, which can facilitate strong affective responses – if not always promoting high levels of critical thinking – at the same time can, however, also have unintended negative consequences around identity formation, especially across online spaces.[4] More positively, Kearney (2011) has shown that learner-generated digital storytelling often leads to the slow development of critical thinking, since it allows learners to express personal emotions through autobiographical explorations of a variety of topics and modes of communication (Gretter *et al.* 2017: 12). Most certainly, digital storytelling can be read as part of an active social practice that challenges, contributes and critiques the inequities and ideological discourses embedded within 'real world' scenarios (Jolls and Wilson 2014). The core line of such scholarship – as explored in the analysis of Al Gore's *An Inconvenient Sequel* explored in Chapter 5 – affirms that in a time of fake news, with the growth of conspiracy theories and edutainment, the necessity for robust forms of interactive and critical media engagement and new forms of environmental literacy becomes more pressing than ever.[5]

Constructing a new consensus around climate change education

Many scholars argue that repeated efforts to shore up the scientific consensus on minimalist claims such as 'humans cause global warming' to help address climate deniers' rhetoric remains a distraction from more urgent matters of knowledge, values, policy framing and public engagement. Consensus quantification is justified, however, by arguing that public ignorance of consensus amongst climate scientists provides a barrier to the implementation of climate change mitigation policy (Anderegg *et al.* 2010, cited in Pearce *et al.* 2017: 1).

In spite of a growing consensus, climate science still remains complex and multi-dimensional, with findings sometimes appearing contradictory. Most importantly the science frequently does not tell civic society anything about what to do to address climate change. Furthermore, as provocatively asserted, 'consensus-seeking is neither a social requisite nor a normative ideal for a viable democracy' (Rescher 1993). At the outset, one could argue, valuing dissensus might encourage and allow a more publicly inclusive and accessible debate over approaches to climate change that do not prematurely foreclose particular policy options (Machin 2013, cited in Pearce *et al.* 2017: 3). All of this is difficult territory to wade through, but emotionally-tuned environmental narratives remain a good forum and outlet for concrete framing and dealing with such (under-articulated) dissensus.

At all stages in the process of improving environmental literacy, it must be stressed that debates about the value of carbon emission reductions cannot be divorced from their social and political contexts (see Cohen *et al.* 1998). All the while, seeking to encourage multimodal and critical engagement with the overall phenomenon remains the overarching goal. However, for some it is easy to be despondent, with an apparent lack of political, much less scholarly, consensus on the topic. Expecting democratic acceptance of the full implications and a worked-out final strategy for dealing with climate change seems unlikely for so many pragmatic reasons, not least the growing levels of inertia and finding it too politically difficult to secure piecemeal acceptance for change, without endless dissensus.

Nonetheless, such a broad-based multimodal approach to address the scale of environmental concerns, using the assistance of media narratives, remains a good first step. As some researchers suggest, this cross-fertilisation would speak to a number of challenges, including the need to:

1. Attend to, and work with, different local meanings of climate and climate change (Hulme 2017) and their relationship to human institutions and behaviour.
2. Negotiate between concerns about the planet as a whole and local expressions of development rights and responsibilities (Jasanoff and Martello 2004).

3 Finding more inclusive ways of fostering innovation within cleaner energy technologies and selecting appropriate levels of investment in climate adaptation.

(Pearce *et al.* 2017: 728)

These and other related approaches and challenges that address key aspects of climate change in particular include: food quality and security; transport and energy; together with probably the biggest nut to crack – namely the move from an (over)consumption-driven economic model of development, towards a more sustainable and low carbon energy future. One hopes all of these can be creatively worked out via the immersive power of audio-visual narratives, assisted by a range of generic modalities. This necessary transformation of society, however, appears at every level incompatible with a normative political/policy strategy of behaviour change. But by encouraging media and communications strategies to actively call on the assistance of a creative imaginative sensibility – across fictional and documentary film, in particular – some breakthroughs can be made on this front. These necessary seismic shifts in global politics, citizenship and creative eco-storytelling also (re)present clear opportunities to connect with apparently disparate issues, human values and policy objectives in more productive ways.

However, all of this requires developing skills in expert judgement across multiple spaces of science, political and public discussion (Raman 2014), rather than simply expecting a singular focus on unilaterally promoting scientific (or political) consensus (cited in Pearce *et al.* 2017).[6] How all of this complexity plays into developing effective environmental literacy strategies and protocols for media studies need to be approached in pragmatically simple ways, while taking into account a range of tensions and other variables. The communication strategy must be designed to ensure all the various layers of concerns are brought into alignment. This difficult process reminds me of the environmental rapper Prince Ea discussed in Chapter 10, who convincingly used a Rubik's Cube to help illustrate the complex nature of environmental tensions that need to be closely re-aligned if a successful model of communication is to be activated.

Designing an effective environmental communications curriculum: a call to arms!

Media literacy as a core pedagogy and educative approach encourages students to actively consider the messages they send and receive, as well as critically assessing all forms of communication. They also need to be encouraged to engage more actively with governmental affairs, appreciate the role of media and messages in general in the construction of their own identities, and more effectively understand the role of values, standpoints, beliefs etc. across their communication choices (Ramsey 2017: 116). By any measure, this is a lot of realignment to cover, much less expect dedicated and well-formed results from sparking an active intervention within the educational system. But a start needs to be made.

At the outset, the requirement of formal education of students with regards to media literacy, according to the National Association for Media Literacy Education, 'recognises that media are a part of culture and function as agents of socialization' (2007: 5). Coincidentally or not, this mode of enculturation also closely aligns with the basic premise of environmental literacy, namely recognising the impact of media/public messages on all processes of socialisation (Ramsey 2017: 124), all the time taking into account all forms of (re)presentations of nature while explaining the growth and proliferation of mass media communication, including the affordances of new forms of interactivity engendered by new media.

There certainly remains an urgent need to promote environmental (and general health) literacy across all aspects of media and communications education. See, for instance, Hargittai (2005,[7] 2009) who developed, validated and updated an instrument to measure people's digital, and more specifically highlighting their web-oriented literacy. Other practical examples include the work of Primack *et al.* (2006), who created a scale measuring adolescents' media literacy with regards to pro-smoking media messages, or Wade *et al.* (2003), who evaluated the link between media literacy and eating disorder risk factors. All of these aspects of media health research, coupled with other related pedagogical measures and techniques – following some tweaking and re-purposing – could be actively used to help embed environmental literacy into a broad-based teaching and research environment.

Media literacy competency generally refers to 'the ability to critically analyse and reflect about media messages, as well as to create and disseminate media messages and take action' (Tulodziecki 2012: 50). At the outset, as Mathea Simons *et al.* asserts, if teachers are to provide their learners with effective media (alongside environmental) education, they should: (a) be sufficiently media literate themselves; and (b) have the required competencies to promote media literacy among learners. By any measure, the notion of critical media literacies should inspire educators to reconsider what it means to be knowledge-holders in society and in turn to value the forms of literacy possessed by students (Song 2017: 67). Such valuation and working with student's interests, while taking into account their media consumption, is an essential first step towards promoting effective environmental citizenship.

For example, the popular technique of using autoethnographies as a research and pedagogical strategy, especially within identity and gender studies, is believed to encourage compelling multimedia interrogations of student's selfhood and community generally. There exists a long-held practice of constructing and sharing autobiographical narratives across media and cultural studies, but often such stories have been framed around critical analysis, rather than as forms of effective media literacies (Camangian 2010). At the other extreme, in his examination of avatars in online spaces, Rafi Santo (2013) has noted that participation in three-dimensional virtual worlds, such as *Second Life*, 'ignited the agentive potential of users' (in Song 2017: 72). In the chapter on video games in this volume, we explored the inherent power of this interactive format to promote new avenues for environmental engagement and enable increased levels of literacy to flourish. Much more work is needed, however, to create pragmatic learning outcomes, using environmental games.

Moreover, Rebecca Black's (2007) examination of the website www.fanfiction.net revealed that interactive virtual platforms help to legitimise young writers as storytellers with powerful alternative identities. This study found that students' reformulations of identity constructions in virtual spaces encouraged them to use their multi-literacy capabilities in order to challenge mainstream narratives and become active members of social causes.[8] Consequently, there are many avenues one could follow towards understanding and addressing the continuum of interactive engagement with new media.

New creative imaginaries and interactive learning

Innovative strategies are always needed to speak to new audiences and call upon progressive aesthetic formats and alternate voices. We need more and more 'creative imaginaries' to help get environmental message(s) across, beyond the clichés of polar bears on ice-flows and factories emitting toxic fumes, while accepting that such well-worn aesthetic strategies and archetypical images continue to have a place in communicating the message. Probably, even if we hoped, there is no silver bullet or one preferred way to communicate the broad range of environmental messages. Nevertheless, audiences constantly need new and innovative ways that connect with the accelerating affordances of new media platforms and marshal the ever-expanding outlets to communicate the global challenge.

Such debates and strategies of engagement need to be radically re-examined, while re-imagining the possibility of more creative imaginaries and drawing on commercial research around mediated objective correlatives that effectively speak to a range of environmental problems beyond the short-hand use of iconic signifiers like those cited above. At the same time, we also need to learn from pedagogical best practices across other forms of educational and (new) mass media literacy, to help find appropriate ways to address the future of successful environmental communication. We certainly need more practical and bespoke media exemplars to address related environmental issues and towards formulating new forms of media literacy that speak to a diverse range of publics and citizens across the world.

A simple but useful formula for instance to deploy might involve the 'Sustainable Human and Environmental Systems' (SHES) approach to education that prioritises 'big picture' thinking and helps students gain a deeper understanding of the overall complexity of environmental issues. As outlined by Reiter *et al.*, the five elements involved include: a systematic approach to learning; systems thinking; revealing complexity; holism; and supradisciplinarity (Burkholder 2017: 6). Pedagogically focused environmental scholars need to use all their undoubted skills in this regard, so that this and other helpful processes can be practically applied to supporting environmental education and actively promoting overall curricular development. Learning through action and through first-hand experience of natural systems is essential as a designated first step towards promoting best practice across all regions of the world.

As also argued by many experts across a number of chapters in this volume, 'neither stewardship of the natural environment nor long-term improvement of the human condition can be achieved without a holistic, integrated synthesis of disciplines, aimed at managing the interaction of both human and ecological systems at their interface' (Reiter *et al.* 2012: 30). Furthermore, as suggested by Vega-Marcote *et al.*, '[I]t is not enough to acquire concepts; it is necessary to learn to put them into action, integrate them and use them adequately under different real-life circumstances'. By all accounts, mere transmission of content and knowledge to students, even if they were receptive, would fall well short of this praxis-based expectation (in Burkholder 2017: 7).[9]

Echoing the long disputed scientific deficit model of transmitting learning, such one-sided attempts at top-down transmission would remain a poor substitute for open-ended and holistic learning. Nonetheless in recent times, it would appear that some level of top down environmental knowledge dissemination remains more necessary than ever, as the general level of discursive engagement is pitched from such a low base. This is illustrated for example through a critical reading of Al Gore's sequel documentary in Chapter 5 and in recognising the need to go back to basics – at least in American media-politics – before beginning to kick-start some advanced form of critical proactive engagement. All the while we must acknowledge the primary necessity for addressing the complex dialogical nature of human learning through critical engagement.

Furthermore, looking, for instance, at how YouTube's promotion of short films, games and music videos – explored in Chapter 10 – impacts on young people's storytelling in distinct ways is hopeful. Such online media, in turn, relate to the particular affordances and active engagement, all of which remains a first step towards a closer understanding of active prosumers of such critical media practice. Millard (2005), for instance, emphasises how young people's (children) understanding of 'narrative forms in different modalities may support one another' (162). Reid (2003) argues that film (alongside all new media narrative formats) can be used to help scaffold effective writing and critical engagement and that students can learn about environmental narratives by 'shuttling' between the two forms of film and print in order to engage with the differing modes and affordances of each form. Reid usefully proposes that print and film studied together can help to make explicit what they have in common and highlight what is specific to each format. Such erstwhile old school adaptation, alongside multi-modal linkages and calling on transmedia protocols, can certainly be extended into the ever-expanding cocktail of new media formats. Unfortunately, a majority of environmental studies of media only focus on print news, primarily because of its research adaptability in affording large corpuses of primary text that can be easily harvested and evaluated using big data software like Lexus Nexus. Most certainly, the educational academy needs more substantial and cross audio-visual media analysis to help capture and reflect the pervasive interest and growing audio-visual media consumption patterns of new generational audiences, who are certainly not limited to written text in their general media consumption patterns.

Furthermore, actually making proactive stories through new media formats contributes to young adults being able to develop and extend their active engagement with environmental storylines, especially across a range of audio-visual formats. According to scholars like Parry, a great deal more thought needs to be given to supporting children at all levels of formal and informal education, towards pedagogically moving between storytelling modes and in developing transferrable skills across the range of forms of expression (Parry 2010: 70).

Somewhat surprisingly however, the findings from Oystein Gilje's 2010 study of multimodal redesign in film-making practices, indicate that students are not adequately capable of transferring particular meanings from the written mode into the language of moving images. This cautionary warning inplies that students often downplay the role of the semiotic tools available to them in the educational context. This lack of easy transferability and translation between so many new media formats and protocols can lead to confusion and even outright disconnect if not downright rejection. Meanwhile, educators are led to believe that new technology is some form of panacea, while being posited as intuitive for young users, especially across so many platforms from Twitter to Facebook, alongside video games and many other formats in between. First-hand experience of teaching media alternatively suggests that there are growing levels of confusion and a lack of joined-up thinking or clear direction applied across much pedagogical practice. Such confusion rubs off on young students, while critically documenting their bottom-up actual consumption and learning from much cross-media output, which ought to be a guiding first principle. This approach helps in providing an impetus towards refining pedagogical practice in the future.

Several ongoing research projects, including Gilje's, also explore how such pedagogical sticking points have raised issues concerning multimodal text-making practices: (a) focusing on how the student deploys and adopts semiotic tools, like script and storyboard and in using peer students as a primary resource in their filmmaking; and (b) how the key scene in a movie could be redesigned or even re-mashed within this filmmaking practice (Gilje 2010: 515).[10] Such pedagogical processes and dealing with practice-based issues need to be further analysed to help create effective long-term teaching practice, while taking into account how the broad range of new media protocols might assist in re-calibrating active media engagement.

At the same time, McLean and Rowsell forcefully acknowledge the pedagogical value of using digital tools and technologies as a way to tap into students' diverse webs of a latent range of literacies and help them make the necessary shift across more academic and critical registers, incorporating various multi-literacies. They make the argument that 'when taken as a visual concept', photography or film for instance allows 'students to shift the way they think about narrative (e.g. the storyboard of plot, characters, settings, point-of-view), to the way they think about photography (e.g. visual tropes and metaphors, angles, lighting, vectors)' (McLean and Rowsell 2015: 104). Of course, this form of visual transformation in learning needs to be re-calibrated further to take account of aural, audio-visual and other new forms of interactive media. Hence, there is much work to be done in the future.

Effective climate change communication: a case study of Earth Institute, Ecomedia, Columbia University

Research and experience, as already suggested, affirms that fear-based arguments have run their course as effective tools for inspiring (environmental) action (Corner *et al.* 2014). To connect with audiences and achieve success within climate change communication, it is strongly affirmed that all levels of communicators need to shift their approach to become more accommodating, especially in promoting dialogical interactivity. Communicators need to go beyond simply providing people with the bare facts about climate change. They need to connect with people's values and worldviews and by putting solutions at the forefront, to make climate change personally relevant to the general public by echoing those values and communities they love. All the audio-visual media and popular narratives explored in this volume, and future contributions that are in the pipeline, seek to make these interactive dialogical attributes come alive for audiences.

Getting the tenor of climate communication right is becoming increasingly important for at least three reasons, according to a case study by the comprehensive Columbia's Earth Institute report: first, the issue itself and the timing of any campaign remains critical. The impacts of climate change are accelerating and delaying meaningful action to reduce carbon emissions increases the probability of harmful impacts. Second, climate change remains abstract, remote, and distant for many people, most of whom are focused on their more personal immediate needs. Third, influential political and economic actors are organising solidly against actions to reduce the carbon emissions driving climate change (*Columbia Earth Report*: 2). All three areas have to be continuously addressed and monitored, as has been stated across this volume, if successful developments are to be carried out into the future.

By any measure, as is frequently asserted across the scholarly literature, communication around climate cannot be a one-size-fits-all exercise. Consequently, there is a continuing need for several related and layered strategies to help promote effective forms of environmental literacy and initiate sustainable changes and long-term transformation. Like all communication strategies, the report lays down a number of markers, while focusing on general first principles, including putting people first and putting yourself as an effective communicator in your audience's shoes. This empathetic communication strategy remains an essential prerequisite for all good pedagogical practice. People interpret new information through the lens of their past experience, knowledge and social context, so naturally this approach needs to be factored into any proposed solution. It must be recognised that different individuals often come to vastly different conclusions about climate change. In part this is because they hold different core values, be they political, philosophical or ethical beliefs. Hence the power of universal narratives, which speak to all aspects of environmental communication – while accepting slippages in meaning and often disparate levels of engagement, depending on a broad range of intrinsic and extrinsic factors – can over time help viewers to empathise and perceive in a more nuanced manner. Certainly, new perspectives around understanding and dealing with complex environmental issues can be brought into the foreground.

As Patrick Murphy (2017) memorably asserts, the dominant way entertainment media present ideas and information about the environment, seem to place the canary, not the mine itself, as the centre of attention. What the short life of the American televisual phenomenon 'Planet Green', for instance, appears to teach us, he suggests, is that despite the best intentions of corporate media towards developing eco-conscious programming and environmental literacy, the details about 'the conflict' and 'the characters themselves will always demand more attention than the looming problems of the commons' (2017: 149). In the growing back-catalogue of quality streamed television series (alongside the growth of other media), some of which has been examined in this volume, the need for new and more developed environmental imaginaries rings out loud. There is certainly much to be done in greening the global media landscape.

Climate communicators most certainly above all else should appeal to values held by their target audience to make it easier for audience members to recognise climate change as a personally meaningful issue. Furthermore, scholars must always recognise that people's response to climate change (and messages about it) are radically influenced by their worldviews, including their sets of deeply held beliefs and attitudes about how the world works and how people should relate to one another (Nisbet *et al.* 2018).[11] Hence, it is well recognised within all communications strategies that stories and media messages have to be contingent, dialogical and evolutionary, while at their core they ought to be consistently robust. At the same time, such attributes also need to be imagined within the confines of concrete strategies being constructed to tell an engaging story around climate change and actively work across a broad range of audience heuristics. Such narratives should always try to get audiences engaged and hopefully, in turn, help to engender a necessary transformation or tipping point on the ground, by at least suggesting, or actively promoting, new behaviour practices that ultimately facilitate a transition to a low carbon and generally more sustainable future.

Coda: online media, data analytics and environmental audience research

Suzanna Priest's (2016) political perspective on science/environmental communication is certainly refreshing in a field that has been dominated by socio-psychological approaches. Her viewpoint calls researchers and practitioners to include more layers of analysis that illuminate how 'scientific truth is distilled through collective processes' (121). To understand such complexity, she skilfully connects the institutional, technological, and ethical dimensions that constitute the social ecology of science communication, while recounting the challenges and opportunities for the field. Priest envisions a critical form of science literacy that provides robust interpretive cues to understand the methodological diversity and uncertainty, which remains intrinsic to the scientific enterprise. While a primary motor in this approach involves looking at both face-to-face and mediated interpersonal communication strategies, at the same time scholars must understand the range of complex political structures which foster or hamper decision-making on climate policies (3).

While being at the other end of the spectrum to a gated public sphere, with the support and protection of regulated quality material and content being vetted and affirmed, potential online audiences can alternatively randomly assess a broad spectrum of non-curated material across the digital expanse of the YouTube platform. For example, by using database search engines and data-analytical protocols, scholars can learn to produce aggregate lists of the most 'popular' environmentally themed narratives, as determined and dominated by the Google search engine – all of which appears to infer that this process constitutes a measure of objective analysis and affording scientifically justified and representative samples of what exists across such sites. While scholars cannot test the accuracy of such analytical datasets – unless blessed with extensive computer skills – nonetheless they can use these metrics as an indicator of overall user popularity, while applying caveats to help maintain their academic scholarly independence and objectivity.[12] Much can be done at the academic coalface through this ongoing dance with data analytics, as a constituent part of future audience research projects.

Many experts still appear to believe that the de facto globalised archive database of Google, alongside other networks and archive banks like YouTube, will remain the dominant mode for the near future at least. These digital archives with their software tools are being deployed to explain everything interesting about this current era, from social to military formations, alongside mapping global capital formations coupled with local resistance. Most certainly, such networks embody 'glocal' combinations by condensing complex clouds of interactions into definite traceable lines of connection (Chun Kyong 2016: 2).[13] The long-term importance of YouTube, or for that matter Netflix, both as a platform and broadcaster and gateway to a 'common culture', remains however very much open to contestation and by all accounts their legacy may be modified in the future. Nonetheless, one cannot underestimate the global potency of YouTube, Facebook, Google, Apple, Twitter and a host of other new media platforms and networks in framing and speaking to a range of environmental issues and their importance should never be underestimated across the full range of media outlets.

Notes

1 As already suggested, recent statistics show that more than 94% of tweens and teens use digital media on a daily basis, including watching online videos, reading, gaming, and interacting on social media, while 34% of tweens and 28% of teens write stories, articles or blogs themselves (Rideout 2015).

Such increasing saturation of new media is borne out by pilot straw poles of my own students in Ireland alongside speaking with international students, anecdotally regarding the ongoing over-saturation of information and 24-hour news cycles. While sourcing some books in our Dublin City University library, I encountered a new check-out system where users can simply stack the books being withdrawn on a base reader. The new surveillance-like machine reads the book's unique bar codes almost instantaneously and places them on the user's account. This is reminiscent of the automated proliferation of so many shopping experiences and is now mainstream across 'digital' libraries. The only problem for the user is having the time and energy to actually read and digest such material.

I Dream of Jeannie (1965–1970) comes to mind for some reason – a much beloved American television (139 episodes) fantasy series, broadcast well before when most of my students were even born. Jeannie (Barbara Eden) had magical powers and could literally flick through a book and digest all its contents in an instant. If only modern-day academics and students with the aid of digital technology could afford similar potentialities? Yet, of course, the promise of the internet and especially Google, alongside new forms of information and digital media literacy, appears to suggest just that – instantaneous information and easily digested 'knowledge'.

2 For instance, science fiction franchises like *Star Wars*, *Blade Runner* and *Terminator* highlight the unique attribute of human beings, as constituted by memories and a biography, which distinguishes them from replicant cyborgs (see Brereton 2005).

3 This recalls the strategies of mainstream and niche media in spreading viral videos.

4 Fanfiction, online advertising and radicalisation provide examples of how young individuals face challenges in determining the blurred line between fact and fiction. While these challenges are significant, educators should not shy away from using the affordances of digital storytelling for teaching (Gretter *et al.* 2017: 10).

5 'Today we move from teaching *with* media and technology to teaching *about* media and technology' (Tiede, Grafe and Hobbs 2015 in Gretter *et al.* 2017: 13).

6 In short, we need the skills for developing and deploying expert judgement in practical contexts, rather than quantitative techniques for capturing consensus in climate science and then using such metrics as a rhetorical driver of climate policy (Pearce *et al.* 2017: 6).

7 Available online at: www.webuse.org/pdf/Hargittai-SurveyMeasures2005.pdf See online paper from Social Science Computer Review 'Survey of Web-oriented Digital Literacy' (2005).

8 This point about civic engagement brings the focus back to Rafi Santo (2013), who has outlined three stages of critical media literacy movements:

- critical literacy – questioning of media biases.
- Participatory literacy – looking at larger social cultures that incorporated media literacies
- Hacker literacies – individuals felt empowered through acts of criticality and participation.

(Song 2017: 73)

9 Some interesting case study students like Janet [not her real name of course] who learned a variety of transferrable skills, but who barely changed their lifestyle in order to decrease their carbon footprint. This failure to act according to the knowledge one

possesses is a common response to the cognitive dissonance that climate change literacy induces (Burkholder *et al.* 2017: 19).

Stoknes describes the state of 'cognitive dissonance' that students can find themselves in: if students are made aware of the dangers of climate change but then are not given meaningful ways to become a real part of the solution, a lack of hope can set in, leading to pessimism regarding the outcomes of climate change action and even a psychologically-defensive rejection of climate change (Burkholder *et al.* 2017: 27).

10 In the case of the Cindy Sherman project, students had varying degrees of reliance on or comfort with technology and digital tools. Many of the 40 students who completed the project wrote in their journals about the frustration they felt not being able to use their cell phones or a digital camera and not being able to edit photos in Adobe Photoshop, iPhoto or some other photographic software. They initially found the task frustrating because it was 'easier to copy and paste' Google images or take photos with their cell phones and edit them on the computer to get the desired effect (in McLean and Rowsell 2015: 110).

11 See this excellent new three volume *Oxford Encyclopedia of Climate Change Communication* (Nisbet *et al.* 2018), which goes to great lengths to recognise communications' 'complexity, taking practical, experimental and creative approaches to knowledge co-production; and continually working to transform power and promote justice' (546).

12 Data analytics are both a blessing and a concern, of course, since we cannot ever be sure if we are accessing an 'objective' representative sample of hits, or simply are pandering to our own pre-given marketing choices and mindsets. At the same time, one wonders what do we miss if we assume new media are simply viral or disruptive?

> Habitual New Media protocols counter this trend to analyse the present through soothsaying by revealing that our media matter most when they seem not to matter at all, that is, when they have moved from the new to the habitual.... Through habits users become their machines: they stream, update, capture, upload, share, grind, link, verify, map, save, trash, and troll. Repetition breeds expertise, even as it breeds boredom.
>
> (Chun 2016: 1)

YouTube media certainly speaks to this two-edged sword.

13 Almost all media scholars talk of how neoliberalism thrives on crisis: it makes crises ordinary.

> Crises are central to habit change: in the words of Milton Friedman, creator of crises par excellence, 'only a crisis – actual or perceived – produces real change.... Information is not Ebola, but instead the common cold. Habit + Crisis = Update.... New media are N(YOU) media; new media are a function of YOU. New media relentlessly emphasise you: YouTube.com; What's on your mind?; You are the Person of the Year.
>
> (Chun 2016: 3)

References

Black, R. 'Digital Design: English Language Learners and Reader Reviews in Online Fiction'. In M. Knobel and C. Lankshear (eds) *A New Literacies Sampler*. New York: Peter Lang. 2007

Bott, C. 'The Internet as a Terrorist Tool for Recruitment and Radicalisation of Youth'. DHS Report, 2009. Available online at: www.homelandsecurity.org/docs/reports/internet radicalisation.

Boyd, B. *On the Origin of Stories: Evolution, Cognition and Fiction*. Cambridge MA: Harvard University Press. 2009.

Brereton, P. *Hollywood Utopia: Ecology in Contemporary American Cinema*. Bristol UK: Intellect Press. 2005.

Burkholder, K., Devereaux, J., Grady, C., Solitro, M. and Mooney, S. 'Longitudinal Study of the Impacts of a Climate Change Curriculum on Undergraduate Student Learning: Initial Results' *Sustainability*, 9(6): 913. 2017.

Camangian, P. 'Starting with Self: Teaching Autoethnography to Foster Critically Caring Literacies'. 2010. Available online at: https://cxarchive.gseis.ucla.edu/partnerships-grants/tiip/tiip-2-workspaces/manual-arts-workspace/libraries/documents-library/Camangian_Autoethnography_2010.pdf.

Cohen, S., Demeritt, D., Robinson, J. and Rothman, D. 'Climate Change and Sustainable Development: Towards Dialogue' *Global Environment Change*, 8(4): 341–371. 1998.

Columbia Earth Report. Earth Institute at Columbia University, USA. Available online at: http://earth.columbia.edu/articles/view/1963.

Corner, A., Markowitz, E. and Pidgeon, N. 'Public Engagement with Climate Change: The Role of Human Values' *WIRE's Climate Change*, 5: 411–422. 2014.

Dunaway, F. *Seeing Green: The Use and Abuse of American Environmental Images*. Chicago: University of Chicago Press. 2015.

Fuchs, C. *Internet and Society: Social Theory in the Information Age*. New York: Routledge. 2007.

Gibson, J. J. *The Ecological Approach to Visual Perception*. New York: Houghton Mifflin. 1979.

Gilje, O. 'Multimodal Redesign in Filmmaking Practices: An Inquiry of Young Filmmakers' Deployment of Semiotic Tools in their Filmmaking Practice' *Written Communication*, 27(4): 494–522. 2010.

Green, M. C. and Brock, T. C. 'The Role of Transportation in the Persuasiveness of Public Narratives' *Journal of Personality and Social Psychology*, 79: 701–721. 2000.

Greenfield, S. *Mind Change: How Digital Technologies are Leaving Their Mark on Our Brains*. London: Random House. 2015.

Gretter, S., Yaday, A. and Gleason, B. 'Walking the Line between Reality and Fiction in Online Spaces: Understanding the Effects of Narrative Transportation' *Journal of Media Literacy Education*, 9(1): 1–21. 2017.

Hargittai, E. 'Survey, Measures of Web-Oriented Digital Literacy' *Social Science Computer Review*. 2005.

Hargittai, E. 'An Update on Survey Measuring of Web Oriented Digital Literacy' *Social Science Computer Review*. 2009.

Heath, S. 'What No Bed Time Story Means' *Literacy: Major Themes in Education*, 1–168. 2004.

Hulme, M. *Weathered: Cultures of Climate*. London: Sage. 2017.

Jarvis, C. 'Fiction, Empathy and Lifelong Learning' *International Journal of Lifelong Learning*, 31(6): 743–758. 2012.

Jasanoff, S. and Martello, M. L. (eds) *Earthly Politics: Local and Global in Environmental Governance*. Cambridge, MA: MIT Press. 2004.

Jenkins, H. 'Transmedia Storytelling and Entertainment: An Annotated Syllabus' *Continuum: Journal of Media and Cultural Studies*, 24(6): 943–958. 2010.

Jolls, T. and Wilson, C. 'The Core Concepts: Fundamental to Media Literacy Yesterday, Today and Tomorrow' *Journal of Media Literacy Education*, 6(2): 68–78. 2014.

Kearney, M. 'A Learning Design for Student-generated Digital Storytelling' *Learning Media and Technology*. 2011.

Kyong, C. and Hui, W. *Updating to Remain the Same: Habitual New Media*. Cambridge MA: MIT Press. 2016.

Law, A. 'Post Truth and Fake News' *MEJ*. 2017.

Lundby, K. (ed.) *Digital Storytelling, Mediatized Stories: Self-Representations in New Media*. New York: Peter Lang. 2008.

Marsh, R., Hicks, J. and Cook, G. 'Task Inferences from Prospective Memories Covaries with Contextual Associations of Fulfilling them' *Memory and Cognition*, 34(5): 1037–1045. 2006.

McLean, C. and Rowsell, J. 'Imaging Writing Futures: Photography, Writing, and Technology' *Reading and Writing Quarterly: Overcoming Learning Difficulties*, 31(2): 102–118. 2015.

Millard, E. 'Writing for Heroes and Villains: Fusing Children's Knowledge about Popular Fantasy Texts with School-based Literacy Requirement'. In Evans, J. (ed.) *Literacy Moves On*. Portsmouth NH: Heineman Press. 2005.

Murphy, Patrick D. *The Media Commons: Globalisation and Environmental Discourses*. Urbana, Chicago: University of Illinois Press. 2017.

Nisbet, M., Schaffer, M., Markowitz, E., Thaker, J., Ho, S. and O'Neill, S. *Oxford Encyclopedia of Climate Change Communication* (three volumes). 2018.

O'Callaghan, P. 'Political Will is a Renewable Resource' Review of Al Gore's Sequel and Interview. *ExBerliner*. 2017.

Parry, B. 'Moving Stories: Exploring Children's Uses of Median in Their Storytelling and the Implications for Teaching about Narratives in Schools' *English Teaching: Practice and Critique*, 9(1): 58–72. 2010.

Pearce, W., Reiner Grundmann, R., Hulme, M., Raman, S., Kershaw, E.H. and Tsouvalis, J. 'Beyond Counting Climate Consensus' *Environmental Communication*, 11(6): 723–730. 2017.

Primack, B., Gold, M. and Switzer, G. 'Development and Validation of a Smoking Media Literacy Scale for Adolescents'. *Arch Pediatr Adolesc Med*, 160(4): 369–374. 2006.

Ramsey, E. M. 'The Basic Course in Communication, Media Literacy and the College Curriculum' *JMLE*, 9(1): 116–128. 2017.

Raman, S. 'Science, Uncertainty and the Normative Question of Epistemic Governance'. In Cloatre, E. and Pickersgill, M. (eds), *Knowledge, Technology and Law: Interrogating the Nexus*. Abingdon: Routledge. 2014.

Reid, M. 'Writing Film: Making Inferences between Viewing and Reading' *Reading*, 37(3): 111–115. 2003.

Rescher, N. *Pluralism: Against the Demand for Consensus*. Oxford: Oxford University Press. 1993.

Reiter, M. A., Focht, W. J., Barresi, P. A., Gill, S., Smardon, R. C. Baker, S. L., Reiter, K. D. and Fitch, E. 'Making Education for Sustainability Work on Campus: The Proposals of the Roundtable on Environmental Systems and Sustainability'. In W. L. Filho (ed.) *Sustainable Development at Universities: New Horizons*, pp. 109–116. Frankfurt am Main, Germany: Peter Lang GmbH. 2012.

Simons, M., Meeus, W. and T'Sas, J. 'Measuring Media Literacy for Media Education: Developing of a Questionnaire for Teachers' Competencies' *JMLE*, 9(1): 99–115. 2017.

Song, A.-Y. 'Critical Media Literacies in the Twenty-First Century: Writing Autoethnographies, Making Connections, and Creating Virtual Identities' *JMLE*, 9(1): 64–78. 2017.

The National Association for Media Literacy 'The Core Principles of Media Literacy Education' *Education Journal of Media Literacy Education*, 9(1): 1–21. 2007.

Tulodziecki, G. 'Approaches to Learning with Media and Media Literacy Education – Trends and Current Situation in Germany'. *The National Association for Media Literacy Education's Journal of Media Literacy Education*, 4(1): 44–60. 2012.

Turner, M. *The Literary Mind. The Origins of Thought and Language*. Oxford and New York: Oxford University Press. 1996.

Vega-Marcote, P., Varela-Losada, M. and Alvarez-Suarez, P. 'Evaluation of an Educational Model Based on the Development of Sustainable Competencies in Basic Teacher training in Spain' *Sustainability*, 7: 2603–2622. 2015.

Wade. T. D., Davidson, S. and O'Shea, J. A. 'A Preliminary Controlled Evaluation of a School-based Media Literacy Program and Self-esteem Program for Reducing Eating Disorder Risk Factors' *International Journal of Eating Disorders,* March 2003.

Zipes, J. *Creative Storytelling: Building Community/Changing Lives*. New York: Routledge. 2013.

Index

For Product Safety Concerns and Information please contact our EU representative GPSR@taylorandfrancis.com Taylor & Francis Verlag GmbH, Kaufingerstraße 24, 80331 München, Germany